T0100724

PRAISE FOR *DIGITAL TRANSFORMATION PAYDAY*

"Digital technologies can be a great compounding investment for any business to enhance product value and position against competitors. However, given the complexity of digital transformation from a technology perspective, it can be challenging to know how it can create value for the organization strategically. But all businesses must develop a level of mastery of digital technologies to remain relevant in the modern economy. *Digital Transformation Payday* offers practical advice and valuable evidence as to how and where specific technologies can positively or negatively relate to value. This book is a must-read for any leader looking to understand how technology investments drive overall enterprise value, based on industry and financial structure, helping guide their digital strategies."

—Vincent Roche, CEO and Chair of the Board of Directors, Analog Devices

"Bottke wrote nothing less than a straightforward and pragmatic user manual for one of the most vital topics of today's business environment: digital transformation and its returns. Once winning arguments for the success of a company, digital landscapes became nothing more than qualifiers. Follow Bottke's framework and turn it into a winning argument again. The new now goes beyond digitalizing of the physical process. The new now is more about softwaring your company. For those who think that time is running out, start reading, it is worth every minute of it!"

—Bram Schot, AO Board Member, Shell plc; Former CEO, Audi; Professor, Practice for Leadership and Transformation, Bocconi University; Senior Advisor, Carlyle Group & Global Cleantech Capital

"Who needs a payback when you are having fun with your design-thinking workshop? Many years into digital transformation, it turns out: We all do. But reaching beyond the individual case study or the qualitative framework is hard work. In *Digital Transformation Payday*, Bottke deep dives into a dataset of over 20,000 observations to get you a full playlist of potential results. Is your digital transformation bound to be 'Money for Nothing' or 'One Moment in Time?' With *Digital Transformation Payday*, you can make sure there is still a chair for you when the music stops."

—Philipp Leutiger, Chief Digital Officer, Holcim

"Bottke substantiates decades of practitioner experience with rigorous academic research. He creates a robust framework that anyone who is in charge of digital transformation at scale will find extremely useful. Part cheat-sheet, part practitioners' guide, *Digital Transformation Payday* is rich in actionable insights that, if implemented in combination, significantly de-risk large-scale digital transformation programs. Most importantly, however, Bottke frames each leg of the transformation journey into the broader corporate value creation narrative: essential reading for any CXO to explain the value of digital transformation up-front, and to reap the rewards once the initiatives come to fruition. "

—Alexander Pavelka, head of Data and Analytics, Red Bull

"Bottke's *Digital Transformation Payday* offers a unique perspective on the promise and perils of digital transformation, blending practitioner experience and insight with extensive research. Gems abound, including a robust distillation of the payday accelerators and decelerators associated with digital technologies (including cloud, analytics, and cyber), getting agile right, and the importance of having a clear and compelling strategy to effectively aim digital transformation. A must-read for those who want to improve the odds of realizing the hoped-for benefits of complex, multidimensional digital transformations."

—Jonathan Goodman, Global Chair, Monitor Deloitte; Vice Chair and Member of the Board, Deloitte Canada

"*Digital Transformation Payday* is an intriguing page turner which looks from all angles at the multiple challenges that exist behind the move to digital. Bottke spells out not only the accelerators of the transformation journey but also the numerous decelerators—in a thought-provoking framework which can drive your future success and enable many digital transformation paydays yet to come."

—Joel Fastenberg, CHRO Asia Pacific, a leading Global Financial Services firm

"I've spent quite a long time on digital transformation projects in the last few years. Very often I've had the impression that returns on such investments were taken for granted more than thoughtfully evaluated. 'What's the real impact on my business if I don't do it?' is one of the questions I've asked myself and my team many times. Bottke's book is particularly insightful in providing answers to the tough questions an executive would—and should!— typically ask. The pandemic forced all of us to look to our businesses in a completely different way and the real question is no longer 'Where do we want to go?' but 'How do we get there?' Digital transformation can be complex, especially for those organizations that still operate in silos—because it requires a systemic approach that involves changing mindset, culture, and processes. This book is a really helpful guide for the tricky journey toward an enormous organizational transformation."

—Walter Susini, Senior Vice President Marketing,
The Coca-Cola Company

"In *Digital Transformation Payday*, Bottke demonstrates some hard-learned truths of digital transformation, like the difficulty for high debt/high dividend companies to transform, based on deep, fact-based, comprehensive research. This is not your typical fluffy 'let's be digital' book. Read it to create real value with digital, not to play around."

—Jaime Rodriguez-Ramos, Operating Director, I-Squared Capital

TIM BOTTKE

DIGITAL
TRANSFORMATION
PAYDAY

NAVIGATE THE HYPE

LOWER THE RISKS

INCREASE RETURN ON INVESTMENTS

Library of Congress Cataloging-in-Publication Data

Names: Bottke, Tim, author.
Title: Digital transformation payday : navigate the hype, lower the risks, increase return on investments / Tim Bottke.
Description: Hoboken, New Jersey : Wiley, [2023] | Includes index.
Identifiers: LCCN 2022029468 (print) | LCCN 2022029469 (ebook) | ISBN 9781119894179 (hardback) | ISBN 9781119894193 (adobe pdf) | ISBN 9781119894186 (epub)
Subjects: LCSH: Organizational change. | Management—Technological innovations. | Information technology—Management.
Classification: LCC HD58.8 .B658 2023 (print) | LCC HD58.8 (ebook) | DDC 658.4/06—dc23/eng/20220711
LC record available at https://lccn.loc.gov/2022029468
LC ebook record available at https://lccn.loc.gov/2022029469

Cover Design: Wiley
Cover Image: © Yuravector/Shutterstock

SKY10036133_092322

To my wife, Miriam, and our daughter, Laura. You matter more than anything else.

Contents

Part IV The Science Behind the Book 153

Practitioner Foreword

We often say that in an increasingly digital society, there's no such thing as an "analog transformation." Every significant business endeavor involves digital technologies. CEOs and boards wrestle with how to grow and thrive in a world that daily sees more technological innovation, more data, more connectivity, more automation, and more fusion between digital and physical realities. In this context, every business transformation is a digital transformation. Moreover, every single CEO and their board must understand how to create and sustain value from digital investments that advance their business ambitions. But how do you manage what you can't measure? Problematically for CEOs and their teams, the connection between digital investments and enterprise value remains fuzzy at best. This is why Tim's *Digital Transformation Payday* research and writing is essential work at a consequential moment for business.

So, here we are. A global economy that is absorbing and being propelled by more and more digitalization every day. CEOs, C-suites, and boards that are investing significant time and capital on transformative digital programs to adapt their companies to this constantly changing economy. And a vast unknown in terms of what efforts actually drive durable growth and advantage. For us as a global management and technology consultancy, these realities were a clear call to action. In early 2021, our firm began a vast research effort to study the track records of global companies, their investments in technology, and the potential impact to enterprise value. That's when we learned of Tim's doctoral research at Bocconi University/SDA Bocconi School of Management. It was a match made in client service heaven—rigorous research and quantitative analysis that stood the scrutiny of academic review combined with perspectives acquired through years of serving clients in the trenches of business. Eureka! A guide to demystify the paths to a digital transformation payday!

Our hope for readers is that you maintain a broad perspective when reading *Digital Transformation Payday*. Consider not only today's operating environment and the portfolio of technologies already described, but an ongoing development with more pervasive digital innovation and tools that can elevate customer experience, form more agile supply networks, heighten employee engagement, and so forth. We are entering a golden era of possibility and business innovation, and digitalization is a new source of capital to be deployed in this era. But the access to that capital and effective deployment of that capital won't be even. Paydays won't be assumed; they will be earned. Select leaders and companies will seize the capital and perform better than their peers. It's a thrilling time for competition and growth. Enjoy the read and make your payday bountiful.

Rich Nanda, US Strategy & Analytics Leader, Deloitte
Nicolai Andersen, CEO Consulting Central Europe, Deloitte

Academic Foreword

When Tim approached me to supervise his doctoral dissertation four years ago, my initial reaction was all but enthusiastic: "Another research on digital transformation. Can't stand one more." These were my first thoughts. In my work as a researcher, I'm usually not excited by topics that represent big managerial hypes, and for sure digital transformation has been—and still is!—one of the most impactful in the past 40 years. I must also admit that I was a bit skeptical about a senior executive of one of the most important global consulting companies, with a longstanding expertise in strategy and technology, who, in my very conventional view, was obviously looking for confirmation to his and his company's beliefs on digital transformation, with a touch of science in it. Then Tim sketched his initial ideas, and I was impressed by his genuine willingness to learn more on such an enormous phenomenon, candidly admitting that his long-term expertise as consultant had not yet provided him with the possibility to answer some big questions that his clients posited. My original reaction turned into: "Let's make it a joint learning journey." In hindsight, this was exactly what happened. The outcome was an outstanding doctoral research developed with a practitioner mind.

The book you now hold in your hands heavily draws upon the original research and represents a truly insightful intellectual and practical contribution for a number of reasons. It advances a novel perspective on the topic, puts it in context so that readers can feel it is relevant to them, and adopts an elaborated attention-catching narrative that makes the reading pleasant and intriguing.

As for the novelty, Tim's effort combines solid scientific method with business acumen in providing the answer to the fundamental question to which every executive and entrepreneur embracing a digital transformation path should respond: Is it really worth the investment?

Now, the smart answer to most managerial issues cannot be but: it depends! However, to make the answer even smarter one must explain on what it depends. And this is exactly what Tim does in this book. With a very innovative approach, designed by analogically associating digital transformation to a chemical reaction, the conceptual framework provided in the following pages identifies a broad set of interconnected elements that makes the value of the investment contingent on a number of external (context) and internal (organizational) factors. Therefore, readers have the opportunity to match the characteristics of their own organizations with those factors and gain insights on how to turn the potential of digital transformation they are designing or managing into financial value.

One effect of the digital revolution is that people are getting used to reading shorter and more scattered pieces. Fragmentation seems to be a value of our times. Consequently, a book appears as an anachronistic object, and reading a full book is an unusual activity for the hustle and bustle of today's lives—even more if it is not a fiction but rather a management one! I was the first reader of the original research content of this book, and now that I read it once more in its final form as a managerial book, I have no fear in saying that the reading is completely rewarding. If one compares this book with the many others devoted to the same topic, the reading experience is completely different: a bold point of view instead of vague slogans, complexity made simple instead of simplicity made complex, managerial insights instead of ready-but-impossible-to-implement recipes. An amazing example of how good research can be the best engine for managerial insights.

For this reason, I'm sure the reader will appreciate each part of it, thereby earning a fine-grained map of the uneven territory of the value of digital transformation.

Gabriele Troilo
Senior professor and associate dean at SDA Bocconi School of Management,
associate professor of marketing at Università Bocconi

Preface

The Quest for Financial Impact: Four Years of Research Behind the Book

The original idea to write this book came to me four years ago in a private conversation with one of my clients, the CEO of a leading listed company. Thanks to a longtime trusted advisory relationship he asked me a very tricky question: "Tim, you know my contract with this company ends in two years and for me this is just a platform to move on to a bigger task. Tell me, am I not better off spending a few millions on a digital incubator, investing in one or the other cool start-up, starting to wear jeans and trainers and 'digital wash' my annual report and internal communications, instead of doing what we both know needs to be done—investing half a billion or more to transform my outdated legacy backend over the next five to 10 years to truly differentiate in the market and make my company more resilient for what is yet to come? On a journey likely to be full of pain, failures, and high cash-outs putting a lot of pressure on our stock price and my nerves while I am still here? Who will thank me for that? Will there be a measurable return on invest and positive stock price impact?"

I should not admit it now as a typical strategy consultant, but I had no answer—at least not one with the necessary facts behind me. This I was not willing to accept. So, I started searching, but still could not find it. I talked to experts in my firm and beyond in my network, and I tried to read every piece of practitioner advice out there. Believe me, there is much more than you can digest. Literally hundreds of books, articles, and presentations on digital transformation. Most are terrible and repetitive or self-refencing, but there are also quite a few good ones. So, in the end, in this complex digital world, where you cannot afford to miss a single great idea, I decided to read most of them to make my own experiences. If you would do the same, I am sure you would share my disappointment, as most of them have one of three major shortcomings.

First, there are many inspiring ideas out there but the advice is mostly based on subjective experience, interviews, and conceptual thinking (Brynjolfsson and McAfee 2014; Gale and Aarons 2017; Mazzone 2014; Raskino and Waller 2015; Rogers 2016; Schallmo et al. 2017) with some traits of an almost religious common belief system. For the followers of this system, recent and soon expected disruptive technological advancements generate significant challenges and opportunities that need to be urgently dealt with. In this declared "Digital Revolution" (Rogers 2016), "Second Machine Age" (Brynjolfsson and McAfee 2014), "4th Industrial Revolution" (Schwab 2017) or "5th Information Revolution" (Nazarov, Fitina, and Juraeva 2019) traditional management practices are supposed to no longer suffice to thrive or even survive (Gale and Aarons 2017; Mazzone 2014; Raskino and Waller 2015). Not surprisingly, all authors immediately propose predominantly prescriptive playbooks to achieve digital leadership. They are often also providing many interesting ideas to help differentiate from the numerous companies that fail doing so (Andal-Ancion, Cartwright, and Yip 2003; Bock et al. 2017; Charan 2016; Davenport and Westerman 2018; Gartner 2018; Kane et al. 2017; Rogers 2016).

Second, and potentially as critical as a shortcoming, the digital transformation recipes are typically inspired by famous digital-born companies, singular transformation examples, and successful digital platform strategies (Bughin and Catlin 2017; Rogers 2016). If ever focusing on the larger bulk of established corporates, they leverage anecdotal case examples and anonymous surveys with subjective company responses (Kane 2016; Kane et al. 2017, 2018), which provide great qualitative insights but obviously also face many admitted limitations (Kane 2019). Opposing concerns that, without a specific context understanding, the digital transformation journey should not be universally recommended for every firm in any context, industry, or size are far and few between (Andriole 2017; Davenport and Westerman 2018).

Third, even though the "planned digital shock to what may be a reasonably functioning system" (Andriole 2017, p. 20) is considered to be "no [. . .] sure salvation" (Davenport and Westerman 2018, p. 3), there seems to be almost no focus on proven financial outcomes for corporates (or any other companies) with the exception of the analysis on EBIT and profit margins by Westerman (Westerman et al. 2014). Quite the opposite; some authors even proclaim that traditional value metrics do not make much sense (Gale and Aarons 2017).

Therefore, and further amplified by the digital transformation surge due to COVID-19, it became very clear to me that there is an urgent business need to develop a more objective, scientifically supported view of digital transformation payback. We needed a new method *to navigate the hype, lower the risks of falling into the digital hype trap, and increase return on investments* beyond the prescriptions of managerial books and practitioner journals. I thus turned to academia and joined a high-ranked global business school as a researcher to gain access to leading-edge academic know-how. With some shock, I quickly realized that science also has so far put inadequate prominence on the digital transformation theme. The few available publications provide only very limited answers. There is little to find of any use other than some initial tries of empirical value/performance analysis, and, therefore, the worry about a "digital paradox, i.e. being unable to capture value from [. . .] digital investments" (Parida, Sjödin, and Reim 2019, p. 14) remains.

In addition, given the turbulent history of multidisciplinary research on technology/IT/IS value impact for capital-market-listed companies, nicely summarized by Kohli and Grover (Kohli and Grover 2008), astonishingly little transfer work has been conducted to reach a similar level of understanding for digital transformation value. The same is regrettably true for the findings of innovation value research (Brockhoff 1999; Salomo, Talke, and Strecker 2008; Strecker 2009; Vartanian 2003) and recent corporate finance research (Damodaran 2013, 2017). Only in few cases do researchers provide some depth and positive results, but, rather, focus on subtopics like observable digital strategy and innovation (Beutel 2018; Mani, Nandkumar, and Bharadwaj 2018) or economy-level entrepreneurship impacts (Galindo-Martín, Castaño-Martínez, and Méndez-Picazo 2019). Others apply qualitative text analysis of statements on digital transformation in voluntary and obligatory reporting (Kawohl and Hupel 2018) and then stop there, or focus on text-analysis-based subjective digital maturities (Zomer, Neely, and Martinez 2018) as a baseline. Except for some recent work linking text analysis and analyst recommendations (Hossnofsky and Junge 2019), and early-stage analysis looking deeper into the firm value and performance implications of non-technology companies investing in digital technologies (Chen and Srinivasan 2019), not once was a broader, end-to-end view of the digital transformation process and its value impact found.

So I had no choice; I needed to embark on a journey by myself to solve this puzzle. To provide answers which, years later, hopefully my client still would want to hear, now that he has moved on and is—no surprise—facing the same question only six times bigger. Based on the focus of my experience and interests, I put my emphasis on the value implications for large-capital market-listed incumbent firms with a sufficient size and history to assume a transversal transformation rather than greenfield digital start-ups or smaller firms. Nevertheless, for this book, I will also discuss implications for smaller companies in a small excursus on this topic in Part II of this book.

I developed a more objective, scientific understanding of digital transformation and its (shareholder) value impact. The main contribution of the research behind this book was the first-ever combination of a structured transversal digital transformation framework with an in-depth empirical analysis of financial impact implications enhanced by a unique, Natural Language Processing (NLP) supported dataset (> 20,000 observations) of large, listed corporates. Good news for you: As you will be able to see later, based on this approach, some advanced statistical analysis, and causal machine learning, there are means to better predict the likelihood for a statistically significant market value payday for your digital transformation efforts, depending on the context of your company.

Navigate the Hype: How to Read
Digital Transformation Payday

Obviously, you can read only parts of this book based on which topics interest you the most (the chapter titles try to tell you what's inside). Still, I believe you will benefit much more if you follow it in sequential order. Like digital transformation, this book is an end-to-end process where every element needs to have its place. However, there is one thing you must be aware of when doing so. What you will not find in this book are explicit company names when describing the practical examples used to illustrate the ideas. This is for a good reason. The payback mechanisms of digital transformation are one of the best-kept secrets, and the likelihood of reading the true story, when this is based on externally sharable information, is close to zero or way behind the fact. Therefore, this book takes a different route and combines up-to-date hard data of what is publicly reported in thousands of annual reports over the past 10 years with a few deep but fully sanitized real-life experiences. Furthermore, you must be aware that this book was developed with the ambition to be scientifically sound (in the end it is based on almost four years of doctoral research). This is meant to end the times of unreferenced rephrasing in digital transformation and implies that there will always be sources and no faked originality by renaming where others before me have already laid the basis in meaningful thoughts.

This book is structured in three parts, 15 chapters, and a series of appendices:

In Part I, *Chapter 1* serves as an introduction. *Chapter 2* will give you more background on the seminal research on technology, IT, innovation, and corporate finance, which, for all the wrong reasons, is so far not reflected in current digital transformation practice. It also argues what your benefits of undoing this mistake are and provides initial learnings of relevance for your transformation programs. We will then dive much deeper into a key and heavily advertised methodology in the market, namely, the numerous concepts of "digital maturity." This section is meant to help you to de-mask this widely misused tool and all its related indexes, benchmarks, and assessments. *Chapter 3* introduces the approach you have been waiting for, digital transformation payday, a new five-element framework enabling you to manage your digital transformation toward accelerated payback by always clearly understanding the complexities of digital transformation benefits and associated cost.

Part II explains the digital transformation payday framework in more detail in eight chapters. *Chapters 4* and *5* elaborate on the complex field of supply- and demand-side digital transformation catalysts, which must be understood far beyond digital technologies. *Chapter 6* raises the issue of transformation scope and the ultimate objective of transforming your core business as a key success factor. *Chapter 7* demystifies agile as a universal tool for any digital transformation process and introduces hybrid as a potential solution to mitigate the downsides of wrongly implemented agile approaches. *Chapter 8* continues by explaining the different subjective and objective angles of measuring digital transformation outcomes to better understand payback and return on invest implications. *Chapter 9* strongly reemphasizes the importance of a strategy to win for your digital transformation success. *Chapter 10* provides two examples of how our framework fits together end-to-end and *Chapter 11* provides a short excursus on implications for smaller companies.

Part III leverages the extensive empirical research underlying this book to help you better recognize the predictors in your business that influence your digital transformation payday and the efforts required to make a difference as a manager, executive, or investor: In *Chapter 12*, the general relationship between digital transformation and value is assessed. *Chapter 13* covers the implications of the industry context you are operating in. *Chapter 14* explores your financial strengths and weaknesses. *Chapter 15* covers the sentiments you create in your capital market communication on what you are doing in terms of digital.

The *conclusion* provides a warning from oversimplification, a high-level payday checklist to keep your paydays coming and an outlook on exponentials—the upcoming next wave of shifts, which will impact you even more in the future than digital transformation currently does.

The appendices add more information on the unique research and data-set behind the book, plus they provide more detailed tables and additional literature recommendations for you to read in case you are interested in diving in even deeper.

Acknowledgments

Many people have been instrumental in creating this book.

First and foremost, I would like to thank my wife, Miriam, and my daughter, Laura, for their love, enthusiasm and motivational support, even when I spent crazy times of the day and night, weekends, and holidays writing or thinking about writing.

Furthermore, I would like to thank my by now faculty colleagues, the great professors and researchers of Bocconi University in Milano, Italy. First and foremost, Gabriele, the best guide I could have asked for, who helped me structure the complex topic, led me back from 80-20 consultant thinking to scientific research, and always inspired me with new angles and ideas. I would also like to thank Dirk for his help on the NLP parts of the research, Andrea for his help on financial valuation models, Alfonso and Andrea for their help on econometrics and statistical models, and Fabrizio for being the Bocconi mastermind in bringing practitioners like me back into the field of research.

From the wider Monitor Deloitte and Deloitte ecosystem, I would like to thank Roger, certainly one of the leading strategists of our time, and Deloitte Executives like Edgar, Jonathan, Nicolai, and Rich who all encouraged and supported me in many different ways over the years of making this research and book happen. I also would like to thank Jaime, my brother in arms in two firms over many years, who brought in his unmatched digital expertise and inspired me with his first book to also start writing. I thank Patrick for his help in structuring the book and making the chapter titles what they are today. Not to forget all the Deloitte and former Deloitte experts, whom I could always ask for guidance on details: Tim (on digital experience and agile), Stephan (on cloud), Stefan (on cyber), Peter (on cognitive and automation) and Jens (on blockchain), and many more. It was always comforting that there is no topic where there would be no expert to talk to to nail the details. Not to forget our great global marketing team, who have worked hard to make sure that this book will be known to and hopefully read by a broad audience. Thank you, Selina, Valerie, Daniela, Sissy, plus Daniel, Florian, and Jens who contributed their creative mind for our campaigning and Diana from our global research team for her unique insights in digital transformation research.

A big thank-you also goes to the initial volunteer readers of this book in its pre-final form. Your unmatched experience in the digital transformation field and constructive feedback and endorsements are very much appreciated.

Finally, I would like to thank the Wiley team, Richard for his trust in supporting my first book based on a rough sketch, my editor, Christina, who enabled me to produce something much better than my own writing while allowing to preserve my own style, and Jessica plus Michael for their support to take the book to market.

PART I

Will Your Digital Transformation (Ever) Pay You Back?

CHAPTER 1

The Bad and Good Reasons for Your Digital Transformation

igital transformation has become an omnipresent buzzword, which is hardly ever fully understood in all its implications. Neither the word *digital* nor the consequences of what it means to really *transform*. Still, in today's business world there seems to be no escape from the pervasive hype. In fact, it is probably one key reason why you have just bought another digital transformation book. And likely this is not your first. But before you dive digital-native-style straight to the deep end, please take a step back and ask yourself: "What exactly lured me into buying this book?" Did the sophisticated marketing campaign, the flashy title, or the carefully formulated cover copy get you? Or was it the assumed expertise of the author, a professor from a top-ranked business school who works with the globally leading digital transformation advisory firm? Perhaps all the digital technologies that vendors, consultants, and specialized media bombard you with daily were mentioned prominently enough as a teaser? Did you therefore have the good feeling that skimming, or—worst-case scenario—reading the book cover to cover will provide you with simple recipes and learnings from even more elaborate and researched insider tales of impressive digital transformation successes or catastrophic failures to succeed in the age of digital disruption?

The Risk of Falling into the Digital-Hype Trap

Depending on your answers to the questions I have just asked, I may have bad news for you: The digital hype has likely already impaired some of your business judgment. By answering any of these questions with a careful yes, you have fallen into the trap of many failed digital transformation journeys.

You are not alone. Many executives have been led astray by the unfortunate belief that the digital transformation trend is different from all the management hypes that came before and that it can be your ultimate savior. You only need to learn the right formula. You might be tempted to throw away the traditional fundamentals for building competitive advantage and generating sustainable financial profits and market value or think that you can use new digital forces to twist them to work for you. This is a transformation, after all. Could that mean it's time to replace those old-fashioned basics with an algorithm? That it's time to disrupt yourself before being disrupted? That a clear winning strategy, enough time for it to unfold and to have a thick skin against failure and resistance are no longer needed? Sounds promising, right?

Formulas that promise a quick fix to these concerns have been around for years in many different forms and flavors. So why do around 70% of digital transformation efforts still fail (Saldanha 2019)? Taking the perspective of a methodology geek, you might argue that this number is mostly based on biased questionnaires designed to convince people they need a new approach to digital transformation. Nevertheless, whether the true number is 50%, 60%, 70%, or more than 80%, it still indicates one major issue that should really worry you deeply as a manager, executive, or investor. With such risky digital transformation portfolios and a high share of invested capital likely wasted without plannable payback, many companies will struggle to generate measurable value, whether in sustainable profits or in market capitalization in their respective marketplace. To prevent this, companies should not leverage digital transformation as an objective in and of itself but, rather, as a management tool. If they do not, it will be difficult for them to secure a relevant return on what they invest. Unfortunately, that also makes it more likely than not that there will be no meaningful digital transformation payoff within the tenure of your current management role or your investment timeframe assuming that you are not working on a timeline above five years or more. This is certainly not a position you want to be in. Short-term glory is not a good reason for embarking on a highly risky digital transformation journey.

The Beneficiaries (Hint: It's Not Necessarily Only You)

The big question is, how can it be that whole sectors, industries, and economies, and so many companies have jumped on a train with such a risky return on investment and payback timing? There are five main drivers of this.

First and foremost, no matter how critical I am of any nonreflective business hype, digital transformation is so much more than that. As you will learn later in this book, there are real and substantial shifts affecting you and everyone around you that must be understood and dealt with end-to-end—that is, strategically, operationally, financially and emotionally. There is no way around it if you want you and your company to become resilient and successful mid- to long term. But that does not mean you should blind-eyed follow digital transformation ideas and recommendations without a detailed understanding of the strategy and economic drivers behind what you are planning to do. All potential digital transformation steps usually have two sides of a coin to consider. On one side is the sometimes proven, sometimes naively hoped for benefits like incremental revenues from cross- and upselling or revenue impacts from faster time-to-market, often with unclear timing and size. On the other side is the effort required to make these benefits happen at scale for the company overall, so that they finally matter in terms of payback. Understanding these sometimes direct and often indirect effects in detail is a lot of work and cannot be achieved easily. Instead, in my work as a consultant, I have seen countless uncoordinated digital transformation projects held together with duct tape and described in consultant-developed presentations. Too often, these so-called transformations lack strategy and have no understanding of complex dependencies. Digital transformation alone is no formula for winning.

Second, there is a whole group of market participants for whom the likelihood of benefitting from digital transformation is much higher (a random subjective guess from experience would be 95%) than for you. In this group you have the massive and growing ecosystem of the so-called hyper-scalers (Amazon, Google, Microsoft, Tencent, Alibaba) and other vendors of digital software and infrastructure solutions across any hidden corner of the value chain. This includes companies that offer customer relationship management (CRM), alternative and virtual realities, (big data) analytics, cloud, automation and artificial intelligence (AI), enterprise resource planning (ERP) (yes, this still exists), cyber, and many more. Not only do these vendors provide solutions, which, from a business and technological perspective, are often admittedly much better than any outdated legacy technology you have already in place, but they will also do whatever it takes to continuously feed the digital transformation hype. This is understandable because it is the single most important reason for their existence; it's their core market value driver. That is why the sector's constantly vendor-switching faceless sales executives are in touch with you whenever they can to make their aggressive quarterly license sales, subscription revenue, or other targets; why you are invited to all these nice physical, virtual, and hybrid events and conferences; and why the media and stock markets celebrate them in a way you can only envy. And why you rightfully can never look at the marketed numbers of their high-level business case claims without a butterfly in your belly. Whatever you do, you should be very aware of the business reasons and expected impact of what you are planning to achieve before you engage them seriously. That is your job and it requires that the business has a sufficient understanding of what these technologies can do. This should never be outsourced to your IT departments and becomes a key skill for any successful business unit. In my work as a consultant, I have seen too many implementations of leading-edge digital solutions, which, lacking clear strategic business understanding, requirements, and outcomes, ended up being mere technology tools, maybe more modern and lower cost than the legacy systems, but not automatically better in terms of supporting strategic business success. A good example are state-of-the-art CRM systems that can end up being implemented as more or less static customer data and lead funnel management solutions not fully integrated with the new or existing systems around them. From a customer perspective, this changes nothing at all in terms of a superior quality customer journey. And here I am not even talking about the question of whether the changes are means to achieve a winning strategy against your competition. Another more operational example are the functional robotic process automation pilots for the few non-niche volume processes that are never scaled to an extent where they make a real difference in the P&L and workforce structure. So remember to always ask the question of what you want to achieve strategically in your business first, before you even consider selecting a single software or a best-of-breed solution and project portfolio.

Third, coming closely behind or hand in hand with these vendors, you have the consultants like me and my firm. In an ideal world you would expect them to balance their original objective of giving neutral business advice on strategic choices with the far-superior growth rates and profit pools that your huge-scale implementation projects can offer. But do these consultants really fight against the temptation of providing business cases that only support what you have decided to do anyway? The right ones certainly do. But you must be very careful to distinguish between advisory as a presales mechanism for technology implementation work and "true north" advisory aiming to find the best strategic solutions for your business, which, in comparison, will always be a niche business model for many consultants. So always ask whether the consultant is truly your neutral advisor or just an attached beachhead for the waiting implementation troops. When you know what to watch out for, you can easily spot this setup. Always analyze:

- Which part of their internal matrix organization dominates your relationship, the customer/sector/industry side (you) or the functional implementation side?
- Do they give you different options from their ecosystem and provide balanced evaluations or are they too closely linked by incentives unknown to you with a single vendor?

- Do they recommend or at least accept strategy and technical implementation to be different questions potentially sourced from different vendors?
- Do their business cases reflect the apparently high likelihood of failure as a buffer or risk premium?

The answers to these questions already indicate how neutral you can expect them to be.

Fourth, do not forget politicians, governments, and public institutions, who happily follow the digital transformation trend and therefore influence how companies transform digitally either by setting the tone for a market or by regulation and subsidies for investments. For them the beauty of including the word *digital* everywhere is that they are suddenly sounding much more modern and that they can claim that it will solve all serious problems they do not really have a good answer for (new pandemics, the climate crisis and lack of natural resources, the divide of rich and poor, the skilled and unskilled, the age pyramid, the shortage of talent and the lack of competitiveness of the country, and many more). And they do not even have to explain what they mean by *digital transformation* in detail, because no one will really try to understand if it ultimately means that some sort of government subsidy is ready to be mobilized. Here you must always ask whether following what is opportune in your given political system will provide true benefits to your business or is rather a communication hype to selectively join for the benefit of your overall positioning as a public relations tool.

Finally, you have an endless crowd of authors. On one end of the spectrum are the self-proclaimed experts and keynote speakers who most often are tied to the consultant business model. On the other end are the people from the practitioner perspective almost invisible researchers from academia, who often struggle to link digital transformation and return on investment through systematic analysis, making much of their work inadequate for the size and impact of the business phenomenon.

In short: There is no simple truth out there.

CHAPTER 2

Why Digital Transformation Advice Can Get You Off the Payback Track

I hope that you agree with me that achieving more certainty on digital transformation payback is an objective worth pursuing. But to do that properly and to judge the quality of any related digital transformation advice, one question must be solved first: What is *digital transformation*?

One would hope that the flood of publications available has converged toward a common definition and understanding. *Ha!* Among practitioners, clear definitions of digital transformation are not of much interest. The elements and frameworks used in this field are as diverse as their inventors' backgrounds. No wonder, since each one is typically affiliated with a professional services firm aiming to fuel—and thus benefit from—the management hype. To stand out and stake their claim, each consultant offers a seemingly unique framework built on different terminologies, elements, and definitions. As a result, there is no common structure, so the actual drivers, scope, and processes that drive transformation remain unclear. At this point, the only common ground seems to be that digital transformation is more than a technology implementation—rather, it is a business transformation triggered by substantial or even disruptive market drivers, and it affects the entire firm's scope of business in fundamental ways. In scientific research, the digital transformation theme turns out to be just as fragmented. Still, academics at least aim to better disentangle the terms *digitization*, *digitalization*, and *digital transformation*, which are otherwise often used synonymously (Henriette et al. 2015; Osmundsen et al. 2018). It is "important to distinguish the three, to obtain a more informed dialogue based on a constant use of terminology" (Osmundsen et al. 2018, p. 3). I could not agree more, as the issue of a common language of digital transformation is one of the initial barriers of transformation success we see in our daily work with our clients (Nanda, Gurumurthy, et al. 2021). More than once, I have seen different vendors working together on a digital transformation program needing months to finally understand that everyone has been working based on different assumptions about what is meant by the different teams, with dire consequences for the progress of the program.

Therefore, I cannot and will not spare you from some clear definitions throughout the book.

- *Digitization*, in its most basic form, is defined as "the process of converting analogue signals into a digital form" (Tilson et al. 2010, p. 749) or, similar, "the conversion of mainly analogue information into the binary language understood by computers" (Hinings et al. 2018, p. 52). This view is narrower than others, which are already blurring boundaries to leverage digital to adapt processes (Verhoef and Bijmolt 2019; Verhoef et al. 2019).

- *Digitalization* refers to the transformation of products or services into digital variants to offer benefits versus the original tangible products (Gassmann et al. 2014).

- *Digital transformation*, on top of being driven by digitalization-related elements, also needs to clarify what it takes to be transformational. It needs a fundamental redesign "by redefining business capabilities and/or internal or external business processes and relationships" organically or inorganically, with the objective to "gain considerable competitive advantage" (Dehning et al. 2003, p. 654). This requires a clear strategy, patience, and resilience against the imminent resistance (Martin 2021).

What You Should Learn from Technology Payback Research

Surprising for me, as a practitioner digging deeper into the wide field of applicable research, there is no lack of relevant empirical findings as such, but, rather, of transfer work from neighboring disciplines (technology/IT/IS value, innovation value, and corporate finance) to digital transformation. You might now ask, why should I bother about such research? Is this not all out of date in the age of digital disruption, as well as far away from real-life applications and either too complex or too simple, depending on from which angle you are looking? Yes and no. Yes, because what is missing is the modernization of concepts to leverage in current times. And no, because the need to understand payback impact of any technology transformation program including digital transformation based on solid research has never been bigger. We cannot afford to leave tens of years of research aside based on the bold (and wrong) hypothesis that the basic rules of technology economics are now void just because we have added the word *digital* all over the place. To dissolve this dangerous arrogance and provide recommendations for practitioners with a scientific foundation are two of the main reasons why this book exists.

The main neglected research field is the decades-spanning work on technology/IT/IS value (Brynjolfsson and Hitt 2000; Devaraj and Kohli 2002; Hitt and Brynjolfsson 1996; Kohli and Devaraj 2003; Kohli, Devaraj, and Ow 2012; Kohli and Grover 2008; Sabherwal and Jeyaraj 2015; Saunders and Brynjolfsson 2016). Almost no other phenomenon in technology has been looked at from more angles and based on more empirical data. Still, as of today, no relevant practitioner or academic systematically builds on the similarities (abstract and technology-related actions aiming to increase firm value) and overlaps (digital as an innovation and as a technology) between digital transformations and the technology-driven transformation research from the past decades. The so-far completely underleveraged work includes early work on technology value impact in general (Clippinger 1955; Morone 1989). It continues with the turbulent and extensive history of analysis on technology/IT/IS value impact, including a long period of IT productivity paradox controversies with heavily questioned value impact, and ends in the widely accepted proof that technology transformation can generate value—not universally, but under certain conditions, according to Kohli and Grover (2008). However, all this research places limited prominence on the defining business drivers in early phases of the transformation process and the process as such and therefore could not provide everything that was needed for the main questions of this book.

Given the symbiotic association of digital transformation to innovation language and concepts (Andal-Ancion, Cartwright, and Yip 2003; Hinings, Gegenhuber, and Greenwood 2018; Nambisan et al. 2017; Scott, Van Reenen, and Zachariadis 2017; Venkatesh and Singhal 2019), findings of innovation value research can offer further input to the following:

- Research on *innovation focus and innovation field orientation* (Henderson and Clark 1990; Salomo, Talke, and Strecker 2008; Strecker 2009) and the idea of the need for a balance of "market orientation," "technology orientation," and "competitor orientation" (Strecker 2009) fit nicely with the later explained catalyst ideas.
- The notion of the relevance of the *distance to the core business* (Strecker 2009) is widely matching with the later described scope definition.

- The extensive *empirical work on innovation-value impact* did not only guide the digital transformation outcome ideas developed for this book but also provides endless inspiration (Brockhoff 1999; Strecker 2009; Vartanian 2003) when designing robust statistical methodologies for analyzing financial value implications (see outcomes described later).

Due to the manifold uncertainties and apparent complexity in any digital transformation, this book also needs to look way beyond classical corporate finance research to capture the outcomes and financial value of such transformations adequately.

This book is heavily inspired by findings on *valuation under beyond-usual uncertainty* (Damodaran 2013) and from situations in which qualitative transformation storytelling (non-accounting "other information") can have a substantial impact on how shareholders see the future of valued companies (Damodaran 2017).

Along the same lines, seminal work on advanced valuation methods from *real options* research (Copeland and Antikarov 2001; Fichman, Keil, and Tiwana 2005; Schneider 2018; Smit and Trigeorgis 2009) is conceptually considered when deemed relevant, mostly to explain the (latent) strategic-option-building idea in many digital transformations.

Furthermore, even though a less event-driven approach was chosen, this research also reflects some thinking of *event study-based valuation research* (Dehning et al. 2004; Dehning, Richardson, and Zmud 2003; Subramani and Walden 2001; Szutowski and Szułczyńska 2017), which over time has investigated the impact of many kinds of capital market announcements and their impact on shareholder value.

Finally and most importantly, the *advancements of residual income valuations models* (Ohlson 2001, 1995) and especially the theoretical concept of "other information" build the major foundation for all financial models applied in this book.

So, what does this mean for me? you might ask. The following six key findings—all quotations in the following list directly taken from the work of Kohli and Grover (2008, pp. 27–28) and reinterpreted for digital transformation—should give you more confidence that digital transformation all by itself does not guarantee value, but at least under certain carefully managed conditions it can generate the payback you are looking for. At the same time, the findings also show that there are no simple recipes to generate value in any tech-driven transformation exercise.

1. **"Technology creates value."** The often-cited IT productivity paradox, which questioned the value contribution of IT in general over many years, is no longer debated because by now numerous studies have "proven a relationship between IT and various aspects of firm value (financial, process-related or perception-related)." The same is true for innovation

payback, which has also been demonstrated with similar tools as technology payback in numerous research projects. The good news for you is that, under the assumption that digital transformation shares at least some similarities with IT and innovation investments, there must be a measurable path toward financial payback, if done right. How to do this is obviously the main topic of the remainder of this book.

2. **Technology "creates value under certain conditions."** IT (and similarly innovation if you follow the corresponding research) does "not create value if they are not part of a larger business value creation exercise." The downside for you from this is that embarking on your digital transformation journey mainly only based on technology will most likely not bring you the benefits you are hoping for. Therefore, you must identify and include all key elements of value creation for digital transformation success. How to do this will be explained in detail later, when introducing a new framework to capture digital transformation end-to-end.

3. **Technology-based value "manifests itself in many ways and on many levels . . ."** IT (and comparably also innovation) "can create value in the form of productivity like other forms of capital, in the form of process improvements, profitability, or consumer surplus. They can be on [the] individual, group, firm, industry level[s]. These include strategy-IT alignment, organizational and process change, process performance, information sharing and many others." Correspondingly, measuring digital transformation payback is for sure an even more complex and multidimensional endeavor, but do not worry, we will come to that and find ways that help you manage this complexity.

4. **Technology-based value "is not the same as technology-based competitive advantage."** This implies that there is a "decisive difference between creating value and creating differential value." Without doubt, digital transformation is therefore not a long-term strategy to win by itself, something we will discuss later in more detail when we talk about the importance of strategy.

5. **Technology-based value "could be latent [and] can also create (real) options to reap benefits at later stages of the development."** This means, as you might have already guessed, that digital transformation will not show benefits as fast as you would like and might even initially only enable you to invest into a platform to build on, when the time comes. We will see how to better predict this and find coping mechanisms later in this book.

6. **The causal relationship of technology investments and payback "is elusive."** According to Kohli and Grover (2008), "it is difficult to fully capture and properly attribute the value generated . . ., often due to the difficult task of gathering relevant data" and not due to the lack of proven

valuation theories and statistical and analytical tools to be applied. Fortunately, *difficult to fully capture* does not mean "impossible," as the remainder of this book will demonstrate.

Digital Maturity and Payback Are Not the Same

So far, the favorite tool of many advisors to prove a need for action and lead toward impact has always been the so-called maturity model. *Maturity model* here simply means that you qualitatively measure the progress of a company (e.g., on a scale from 1 to 10 or in a clearly described morphological box) on its digital transformation journey across many different dimensions that are combined in many different ways in one overall model and usually benchmark your results to a planned target state and/or the expected or current state of your competitors. It comes as no surprise that the same is true for digital transformation. The beauty of these models from a sales and business development perspective is that you can easily define these maturity dimensions and criteria in any way you want and then set the thresholds for "good" to "great" high enough to make sure that there is always a case for change. Either in comparison to status quo or, even more convincing, vis-à-vis competition—which is a nice coincidence—can be freely defined based on the argument of eroding industry barriers in the digital age. I am sure you have already been presented digital transformation maturity models, based on your own or others' subjective assessments. As you can see, I am critical toward these models, if applied with the wrong intentions. Still, they deserve some reflection given their prominence under the wider term of "future readiness" (Weill and Woerner 2018), or most prominently "digital maturity" (Kate et al. 2017; Kane et al. 2018; Kane, Palmer, Kiron et al. 2016; Kane, Palmer, Nguyen Phillips et al. 2016; Westerman 2011, 2013; Westerman, Bonnet, and McAfee 2014;). Consistent with my earlier findings, related technology/IT/IS maturity work apparently did not serve as an explicit inspiration or scientific foundation for any digital maturity work. Existing concepts (Becker et al. 2010; Proença and Borbinha 2016) were completely neglected, even though the number of models and tools grew massively starting in the early 2000s (Becker et al. 2010). This is fortunately no problem, because early digital maturity research shared the same main weakness as later concepts: It is mostly descriptive and therefore of little empirical substance (Becker et al. 2010) to help us with our search for payback acceleration.

Nevertheless, (digital) maturity models can have their uses, if you are able to navigate their complexity to understand what they can do for you and what not. To better grasp the different variations and concepts of the digital maturity idea, clustering them in three different types turned out to be helpful.

Abstract Digital Maturity

The first type I call *abstract digital maturity*. Prominent examples for this type include "digital leadership" (Raskino and Waller 2015) or "reach full potential or even overachieve in [the] digital world" (Gale and Aarons 2017). Sometimes this is referred to as being *future ready*. According to Weill and Woerner, "Future-ready enterprises are able to innovate to engage and satisfy customers while at the same time reducing cost . . . using digital capabilities to transform a traditional enterprise into a top performer in the digital economy" (Weill and Woerner 2018, p. 21). This idea of being future ready is easily stated, but most difficult to validate. These defining characteristics remain mostly abstract. In any case, they provide no empirical evidence and are therefore of little use for what we want to achieve other than inspiring the general discussion. I have still used these ideas in some situations. They can be beneficial when you encounter digital-unsavvy companies or managers in initial discussions. For example, in a smaller regional utility, I talked about abstract future readiness with the CEO and the CFO and confronted them with colorful case examples of what digital platform companies can do for their customers already today. The focus was on omnichannel experience, digital-first interactions, chatbots, and other tools relevant to the utility company. This helped to serve as a wake-up call for them to start doing the real work and move from the abstract understanding of the need to action to the more functional improvement areas.

Functional Digital Maturity

The second type is *functional digital maturity*. This type of digital maturity is specific to limited functional areas, for example, "digital marketing governance maturity" (Chaffey 2010), "smart factory implementation maturity" (Sjödin et al. 2018), "web analytics maturity" (Hamel 2009) and others. Understanding this level of digital maturity can be of value for you in the initial assessment but provides little insight on the overall digital transformation that is necessary for your situation. Nevertheless, these models also have their uses, usually as one qualitative element of very clearly defined optimization programs. I have used them, for example, when providing assessments of the data intelligence department of a large European telecoms company. The maturity of the teams, the data generation and consolidation approach, and the underlying technology infrastructure provided great inputs (but not more) when developing a concept of how to take this area to the next level to support business success.

Aggregated Digital Maturity

The third type, aggregated digital maturity, is the most prominent but unfortunately often also untransparent approach in the market. In the best case, it culminates in simplified and easy-to-explain 2-by-2 matrices with widely differing axis parameters. Consultants love this. Table 2.1 provides a selective summary of prominent digital maturity work, selected for best insights and quality. This should help you to better put all current and future models you might be confronted with into some perspective.

These models at least reflect the assumed and, in practice, clearly observable nonlinearity of digital transformation paths (Remane et al. 2017), which are also key elements of the transformation framework later used in this book. All concepts unfortunately do not offer details on the actual approach to aggregate the list of criteria/dimensions to end up with the explained axes. However, they have come up with inspiring marketing-approved names for the archetypes. But as you can see, there is nothing new, no matter what any advisor or author might tell you.

Most of these models share the same problems. First, digital maturity surveys can never be more than a subjective internal view without a reliable correlation or even causality to externally measurable payback parameters. Out of the vast body of digital maturity work, only Westerman, Bonnet et al. demonstrate their digital maturity findings' (based on digital capability and leadership capability as their axes) positive impact on corresponding payback parameters like revenue generation and profitability (Westerman 2013, 2011; Westerman, Bonnet, and McAfee 2014). I have nevertheless drawn on these types of maturity models, sometimes as a first step to build out an aggregated model that reflects deeper analysis across the overall company. One example is the workshops we do to take management teams to a common understanding of how much progress they need to make in each area. The goal of this process is to provide a cornerstone to support a previously designed strategy to win. The beauty of this is that you can also at least try to put an investment envelope behind each area you have worked on.

Conceptually and practically, under the assumption of efficient capital markets (Fama 1970), external observers of digital transformation processes cannot rely on these abstract maturity models. They can only leverage the specific, measurable, accepted, realistic, terminated (SMART) digital transformation descriptions (Kawohl and Hüpel 2018), which capital market listed firms publish or announce in the wake of their legally obligatory or voluntary external communication. No matter what others might tell you, you can work with digital maturity models and still fail to achieve certainty that more digital maturity will lead you to what we are looking for—payback on your digital transformation efforts in terms of market value or earnings.

TABLE 2.1 Prominent aggregated digital maturity concepts.

Author	Axis Parameters	Quadrants/Archetypes	Quadrant Split	Quantitative Analysis of Digital Transformation Value
Westerman, Bonnet, and McAfee (2014)	• Digital capability • Leadership capability In earlier versions: • Digital intensity • Transformation intensity	• Beginners • Conservatives • Fashionistas • Digital masters/ Digirati	Mean value plus half standard deviation along each axis	Yes, revenue and profitability
Berghaus and Back (2016)	• Cluster analysis based on weighted nine dimensions	• Promote and support • Create and build • Commit to transform • User-centered and elaborated processes • Data-driven enterprise	Not applicable	No
Remane et al. (2017)	• Digital impact • Digital readiness	• Clusters 1–5	Mean	No
Weill and Woerner (2018)	• Customer experience • Operation efficiency	• Silos and complexity • Industrialized • Integrated experience • Future-ready	2/3 along each axis	No
Kane et al. (2018)	Linear	• Early • Developing • Maturing	Not applicable	No

CHAPTER 3

Digital Transformation Payday

A New Framework for Accelerated Payback

Before going broader and deeper into the book, the underlying perspective is of particular importance for digital transformation definition and scope. In the micro category, the individual's impact on the digital transformation is the lowest level of aggregation; it is followed by a project-level perspective (Figure 3.1). At the top is a firm-level perspective, where single digital transformation measures or projects are of subordinate relevance to the transversal digital transformation of the firm. Further, two perspectives comprise the macro category, with the industry/market and the (national) economy perspectives.

The empirical parts of this book focus on the firm and industry aggregation levels. Digital transformation can be externally observed without insider knowledge and only here can capital market value development analysis happen, given the irrelevance of the individual and non-transparency and incompleteness of project-level information from an external perspective. These nevertheless play an implicit role when we discuss our framework and the inherent digital transformation payday accelerators and decelerators in each element (see Part II for those). References to macro industry-/market-level perspectives must obviously be included on several occasions to reflect the need for an overall view on the digital transformation process. This is true especially when discussing the drivers or trigger points for transformation (the later explained supply- and demand-side catalysts). It also serves as a key element of the applied statistical models (via the so-called industry effects).

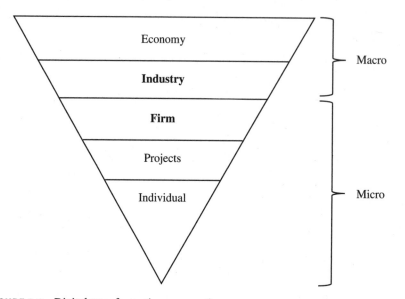

FIGURE 3.1 Digital transformation perspectives.

Given the mess in terms of definitions and frameworks out there, one thing is obvious: A common language is needed to properly manage digital transformation toward accelerated payback. To get closer to this goal, we will use chemical reaction processes as an analogy for digital transformation. On a metaphorical level, these processes and their interpretation allow to better describe and structure potential value implications of digital transformation drivers (that is catalysts), who, "like their chemical counterparts, . . . can amplify and accelerate reactions without being consumed by them" (Briggs 2019, p. 121), the parts of the firm in scope of the digital transformation (that is reactants), the digital transformation planning and management process chosen (that is reaction mechanisms) and resulting digital transformation outcomes (that is products). Very important to add to this concept is that all companies need a strategy (the experimental design) which builds the fundament for any digital transformation to generate a sustainable winning positioning in the market (Nanda, Philips, and Jarmuz 2021). Otherwise, the overall digital transformation experiment will very likely fail or lead to totally unexpected outcomes with potentially disastrous impact, one of which would be that your company has not gained anything vis-à-vis your competitors from the transformation and spent tremendous amounts of money in getting there. While this framework has digital transformations in mind, it is also prepared to consider any (disruptive) technology-driven transformation. This is important as, most likely, the world already is progressing or will soon move to the next stage of development. It is believed to quickly evolve "beyond digital," where so-called "exponentials" (intelligent processes, integrated reality, new energy matrix, digital governance, bioprogramming, and neurogamification) make digital look "boring" (Rodriguez-Ramos 2018, pp. 1–9).

Five Key Elements of Digital Transformation Payback

Digital transformation as we see it (Figure 3.2) is the business transformation of a firm/corporation. These transformations have as explained before five characteristics; they are:

1. Based on a **design/strategy** aiming at new winning ways of doing business built on unique and hard-to-copy real options, new capabilities and/or processes and relationships within the firm or at interfaces to other firms.
2. Triggered proactively or reactively to capture or mitigate already visible or future expected (digital) technology related supply- and demand-side **catalysts/drivers**.
3. Starting with the **reactant/scope** either in the core business, the adjacent business, or at the frontier beyond the existing business and spreading from there to other areas.
4. Planned and implemented sequentially ("waterfall"), "agile," or in "hybrid" **reaction mechanisms/processes** and usually iterative in the sense that after the transformation is before the transformation.
5. Leading on overall firm level to abstract or concrete **products/outcomes**, which ultimately should lead to competitive advantage and measurable payback.

Design/Strategy

Numerous frameworks and strategy approaches have evolved over time, become the hype of the day, and then vanish again (Mintzberg, Ahlstrand, and Lampel 1998). No matter what the approach chosen actually is, one thing I have learned over almost two decades of strategy work is that any transformation approach not based on a clear strategy to win will almost certainly fail. This is probably even truer for digital transformation, where the sweet temptations of digital technology can lead to losing sight of the true goal of any business: sustainable success. Please note that while this element

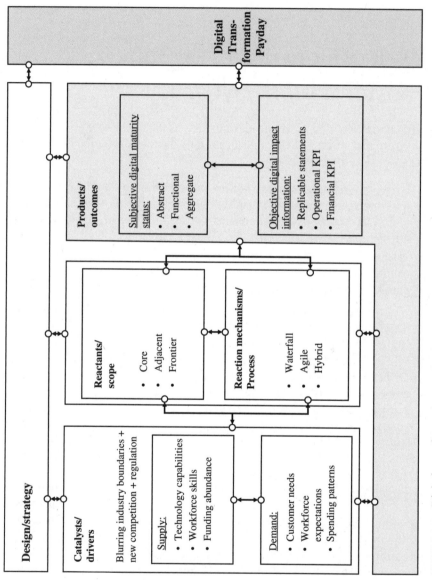

FIGURE 3.2 Digital transformation payday framework.

is obviously the first in our framework, it will still be explained last, as it is from a didactic standpoint easier to understand when you know the other elements first.

Catalysts/Drivers

Organization and strategy research is usually clear in the sense that most if not all strategic transformations of a given firm are influenced by the market context. Numerous frameworks and strategy approaches related to this idea have evolved over time (Mintzberg, Ahlstrand, and Lampel 1998) usually with one idea interwoven: There is a more or less complex supply and a more or less complex demand consideration. In managerial practice, industry sector, industry concentration, or similar data points serve as moderators for statistical firm performance and value analysis, as for example nicely summarized for technology/IT/IS value research by Kohli and Devaraj (2003). For the sake of the book, these categories, in their digital driven combination, proclaimed to enable the "compression of supply and demand" (Gale and Aarons 2017, pp. 34–36), will serve as key catalysts.

Reactants/Scope

As visible in practitioners' and academics' digital transformation work so far, there is some agreement that, in the end, the firm overall is the ultimate reactant in scope of the digital transformation. Beyond prescriptive advice, however, there is no common structure for further breaking down this scope (reactants) in terms of where to start, in which order to proceed, and even less which planning and execution process (reaction mechanism) should be applied to achieve the desired outcomes. As already proven in digital transformation practice and discussed in sporadic managerial articles at least some common themes to better structure digital transformation scope emerge. It all depends where in your firm you start and where you are planning to end. Leveraging—similar to established innovation research— the distance to the core business (Strecker 2009), the distinction between "core," or the "center" business (Gray et al. 2013), the "adjacent" business, and the "frontier" business, also called the "edge" (Gray et al. 2013) is applied to define your starting point or current transformation focus. This is always based on the view that these scope questions only mark the starting point of a digital transformation, never losing the full firm scope out of sight. Because the whole firm needs to adopt (Andal-Ancion, Cartwright, and Yip 2003), firms need to find the right balance between exploitative transformation in the "core" (for example, modernizing their legacy infrastructure) and explorative transformation in "adjacent" and "frontier" areas (for example, launching new business units or diversifying into very different or close-by industries).

Reaction Mechanisms/Process

Interestingly, and opposite to the widely accepted ingredients of a clearly broken-down winning strategy (Lafley and Martin 2013), practitioners are shifting more and more away from discussing the transformation reactants (i.e., the actual transformation scope) to the reaction mechanisms (i.e., the transformation process). Without a clear target in mind, business buzzwords of the day, originated from software and IT development practices, like "agile," "hybrid," "dual-speed," "bimodal," come into play and need to be carefully described and understood to allow for a meaningful integration into our framework.

Products/Outcomes

Finally, outcomes (products) are structured in two different categories with distinct characteristics: abstract outcomes and concrete outcomes. The first category includes all subjective digital transformation outcomes, which are not observable by shareholders. Unfortunately, it still forms the major body of digital transformation impact analysis in research and practice and encompasses all questionnaire/survey or other subjective analysis-based maturity models as elaborated in much detail in Chapter 2. This category serves as an element of the transversal framework for the sake of completeness, but it is of little value for the empirical-driven research objective of this book. Instead, the second, the objective category, likely includes more unbiased descriptions (from now on called "replicable references" which are explained in Chapter 8 in detail as a key concept of the research underlying this book), best proxied by publicly available statements on digital transformation efforts, actions, or qualitative achievements, operational KPIs—that is parameters, for example, on customer experience or employee experience (net promoter scores)—and productivity parameters and financial KPIs—that is directly linked digital transformation data from P&L statements, balance sheets, cash flow statements, and these can provide more adequate inputs for empirical findings.

For your convenience I have added a brief overview of key digital transformation literature from practice and science in Appendix A and clustered these publications into our framework.

While these elements are crucial to better structuring our digital transformation payback thinking from an end-to-end perspective, real-life transformations are often highly complex. Every single one of them always has two sides, both with challenges in estimation, measurement, and management and both with potential impacts on accelerating or decelerating your digital transformation payday.

But how to better grasp this multidimensional complexity? For the sake of this book, I have decided to not make things worse with another cost-benefit framework beyond the dimensions already required for our analysis. Instead, the discussion of each element will be wrapped up with a simple table in four quadrants as shown in Table 3.1.

One dimension lays out whether the item is more an accelerator or a decelerator of your digital transformation payday; the other just separates the more tangible (easier to measure and quantify) and intangible (more challenging to measure and quantify) items.

Simply put, we have on the accelerator side to speed up your payday: More tangible are (1) the easiest to estimate and measure cost savings; more intangible are (2) the more difficult to estimate and measure revenue increases and (3) the often abstract and therefore ignored customer surplus implications with manifold indirect impacts on cost savings and the topline.

However, every digital transformation element also comes at a cost to potentially delay your digital transformation payday. More tangible are (1) the easiest to plan and measure capital expenditures and (2) the plannable and measurable investments in operational expenditures, and more intangible is (3) the often abstract and therefore neglected opportunity cost of embarking on a digital transformation journey. These categories will later help to structure all cost–benefit discussions in simple-to-understand buckets.

TABLE 3.1 Clusters of digital transformation payday levers.

Digital Transformation Payday Levers	More Intangible	More Tangible
Accelerators	• Incremental revenues • Indirect benefits from customer surplus	• Cost savings
Decelerators	• Unplannable cost • Opportunity cost	• Plannable capital expenditures • Plannable operational expense increases

The Path to Payback Is Never as Linear as You Hope

So why is the preceding framework not as nice and sequential as you could expect and maybe even hope for in a business book? Because reality is not sequential.

First, all elements can and will influence each other in many directions. This means that not only do a multitude of catalysts drive your transformation concurrently, which can define which scope (for example, the core in terms of efficiency increases or the frontier by enabling new business models) is affected most but also, in return, the transformation scope can define which catalysts (for example, digital technologies) have the highest importance for the digital transformation process. Another real-life example could be that the applied process is not only a function of the scope and catalysts coming before but also, if the digital transformation is initially focused on implementing agile ways of working, it almost naturally starts in clearly siloed units either at the frontier or in adjacent business units. The same is true for the value backflow from digital transformation outcomes, which in return can influence all transformation elements before them, depending on the actual impact achieved. Cost efficiency from a digital transformation process, for example, can free additional sources for funding as a catalyst for further change, a phenomenon which companies often describe as "save to transform" programs in their internal communications.

Second, even though the described elements do not necessarily always create direct outcomes, they can and often will. This allows measuring outcomes for each element separately via a carefully designed proxy based on externally observable information (using a variable called DIGITALPROXY, which is introduced in the next section).

The Only Econometric Formulas You Need to Understand

You already know and now hopefully continue to read this book for its major objective: to better link digital transformation practices to the value impacts you can later observe in real market and profit and loss (P&L) statement numbers. Consequently, there is no way around some econometrics. But do not worry; what you must understand in detail from the complex underlying scientific research process is quite limited. Two empirically tested formulas are applied to identify statistically relevant relationships of externally observable digital transformation proxies (the earlier mentioned "replicable references") to selected value parameters: The first one is for market capitalization (MARKETCAP), to better understand shareholder returns, and the second one is for three-year average future return on assets (ROA3Y) to better comprehend lagged future earnings.

For market capitalization, the main philosophy behind all financial analysis was to apply a balanced mix of scientific rigor and practitioner-minded pragmatism. This implied no distraction by specialized valuation research models that only academics care about, and it helped to keep track of accessible digital transformation data, instead of dwelling on theoretical peculiarities of detailed model parametrization. Financial valuation methods have been grouped in many ways, depending on their objective and approach (Falkum 2011, Vartanian 2003). Direct and foundational digital transformation-, technology/IT/IS-, innovation- and corporate finance-value research over time has applied seemingly "infinite" variations of customized approaches and standard models out of this basket of potential valuation analysis. From a practitioner perspective, however, further reviewing this substantial breadth of models was of no further interest. Instead, the most promising angle was one subset of approaches, which helped to find sound ways to integrate non-accounting digital transformation proxies as "other information" into models otherwise based on publicly available financial reporting. After a careful assessment, as detailed in Appendix E (for the ones of you who may be interested in digging into deeper), residual income valuation-based models, and specifically the Ohlson model (Ohlson 1995, 2001), seemed to be the best tool for the objectives of this book. Since being introduced by Ohlson in the 1990s (Ohlson 1995), these residual income valuation (RIM) models have opened new possibilities to leverage accounting-based information in valuation. They have been shown to produce better results than the earlier favored cash-based models (Gao et al. 2019; Ohlson 2001). In simple terms, investors are presumed to trade current asset value for a future stream of expected income. Asset prices embody the present value of all future dividends expected. The Ohlson model ". . . replaces the expected value of future dividends with the book value of equity and current earnings . . . based on the . . . clean surplus principle, which holds that the change in book value of equity will be equal to earnings less paid-out dividends and other changes in capital contributions" (Muhanna and Stoel 2010, p. 50). Nevertheless, in its basic form, the model was not without criticism, often with the claim of undervaluation. This is further elaborated and interpreted in Appendices B and E, together with more details on all underlying formulas.

Including a few additions specifically developed for this book, the applied regression to find potential relationships between digital transformation and market capitalization is:

$$MARKETCAP_{jt} = b_0 + b_1 DIGITALPROXY_{jt} + b_2 TOTALEQUITY_{jt}$$
$$+ b_3 NETINCOME_{jt} + b_4 ACOI_{jt} + b_5 PAYMENTOFDIVIDENDS_{jt}$$
$$+ b_6 DELTAEQUITY_{jt} + b_7 REVENUEGROWTH_{jt} + b_8 ROA_{jt-1}$$
$$+ b_9 DELTAEARNINGSDATE_{jt} + b_{10} POLARITY_{jt}$$
$$+ b_{11} SUBJECTIVITY_{jt} + b_{12} BOOKTOMARKET_{jt}$$
$$+ b_{13} INVESTEDCAPITALGROWTH_{jt}$$
$$+ firm | industry + interactions + year + e_{jt}$$

where DIGITALPROXY is a proprietary variable to measure the firm's digital status (see Appendix E), and all other financial variables are self-explanatory for firm (j) in year (t) and year and firm|industry indicate the identified time and firm/industry fixed effects, interactions the possible variable interactions, and $\left(e_{jt} \right)$ is the error term. POLARITY and SUBJECTIVITY are additional sentiment measurements as explained in Appendix D.

While this market value impact of digital transformation is of major interest for this book, it seemed advisable to look at least into one more directly accounting-driven parameter as well. Future earnings were the most obvious choice, given their relevance also in the market value considerations. Instead of developing one's own customized model for future earnings analysis, it seemed best to follow Muhanna and Stoel (2010) and their approach. Other than for the Ohlson model, no extensive empirical tests for such mixed fundamental models exist. Therefore, potential shortcomings were fully addressed with one's own statistical robustness tests. As no logic or empirical results dictated otherwise, a similar approach as selected for residual income regressions was also applied for the mixed fundamental model. It uses the forward average of return on assets (ROA3Y) over three years (current plus the next two) as the measure of future earnings to allow for a possible lag between digital transformation parameters and the realization of potential value. The used regression is:

$$ROA3Y$$
$$= b_0 + b_1 DIGITALPROXY_{jt} + b_2 TOTALASSETS_{jt}$$
$$+ b_3 NETINCOME_{jt} + b_4 NETINCOMEGROWTH_{jt}$$
$$+ b_5 REVENUEGROWTH_{jt} + b_6 POLARITY_{jt}$$
$$+ b_7 SUBJECTIVITY_{jt} + b_8 DELTAEARNINGSDATE_{jt}$$
$$+ firm | industry + interactions + year + e_{jt}$$

where all variables are self-explanatory for firm (j) in year (t) and year and firm|industry indicate the found time and firm/industry fixed effects, interactions the possible variable interactions, and $\left(e_{jt} \right)$ is the error term. POLARITY and SUBJECTIVITY are additional sentiment measurements as explained in Appendix D.

Why do you have to bother with the math? Because, as Appendix E details, these relationships are not only theoretical formulas from the academical world with no real-world impact, but on average they also exist in factual life data in a statically significant way. They even show some initial signs of causality. This means that what you do in terms of digital transformation makes an impact, either directly—measured by the digital transformation proxy—or indirectly, via the financial parameters used to find relationships to market capitalization and future earnings. This was extensively tested for this book. So, whatever you do in terms of digital transformation, you can be more confident that it will make a difference in terms of payback when you start linking in the cost–benefit levers and these formulas.

PART II

Five Key Elements Drive Digital Transformation Payback

CHAPTER 4

Supply-Side Catalysts

Digital Technologies Alone Do Not Do the Trick

Remember, we assume that digital transformation catalysts, "like their chemical counterparts, . . . can amplify and accelerate [digital transformation] reactions without being consumed by them" (Briggs 2019, p. 121).

We structure them in two clusters with three elements each: a supply-side, which will be explained in detail in this chapter, and a demand-side plus overarching elements, which will be explained in the next chapter.

On the **supply**-side we find the following catalysts:

- Technology capabilities
- Workforce skills
- Abundance of funding

Digital Technology Does Not Scale Without New Skills and Funding

Our supply-side catalysts cover all technology-driven transformations and have one thing in common: The relation to or dependency on technology capabilities. But it is not all about technology and spending money on digital is not enough to generate value (Salviotti 2022).

We will also need to discuss other catalysts with often less emphasis but no less relevance. On the supply side, they include workforce skills plus funding mechanisms.

The good news is that no matter how unique and disruptive you think digital transformation may be, there are still commonalities among most extensively researched technology/IT/IS transformations and digital transformation. We can and should benefit from this. Still, the corresponding technology/IT/IS value concepts often do not refer to specific technologies. They rather focus on constructs such as the following:

1. IT investments (Devaraj and Kohli 2002; Kohli, Devaraj, and Ow 2012; Muhanna and Stoel 2010; Sircar, Turnbow, and Bordoloi 2000)
2. IT assets (Aral and Weill 2007)
3. IT capabilities (Bharadwaj 2000; Leonardi 2007; Mithas, Ramasubbu, and Sambamurthy 2011; Muhanna and Stoel 2010; Saunders and Brynjolfsson 2016)

This is because they are often based on resource-based view and dynamic capabilities thinking (Aral and Weill 2007; Bohnsack et al. 2018) and therefore refer to technology mostly as an abstract placeholder. For only a few (McAfee and Brynjolfsson 2008; Peppard and Ward 2005) does a specific technology (for example, enterprise resource planning (ERP) and customer relationship management (CRM)) play a more relevant role.

But digital transformation is much more. It is not just an IT investment or (dynamic) IT capability but a portfolio of new and each potentially disruptive technologies becoming available within much shorter time frames (Kane et al. 2018). It can thus significantly reduce the required energy levels or even create an imperative (Fitzgerald et al. 2013; Gale and Aarons 2017; Raskino and Waller 2015) for transformational change. However, "becoming a digital leader isn't simply a matter of technological savvy" (Baculard 2017, p. 2). In this new world, not a single technology as such becomes a success factor, but rather the capability of a firm to monitor, evaluate, and capture beneficial technology breakthroughs at the right time (Andriole 2017), the "triple tipping point," where the technology, the regulatory environment, and social developments play together (Raskino and Waller 2015, pp. 43–61) for maximum strategic impact. This is in line with innovation value research, clearly demonstrating the positive influence of managing "technology orientation," "market orientation," and "competitor orientation" in an integrated way (Strecker 2009).

Digital Technology Capabilities

There is substantial variation in actual naming of digital technology capabilities (Bohnsack et al. 2018; Briggs 2019; Brynjolfsson and McAfee 2014; Gimpel and Röglinger 2015; Parida, Sjödin, and Reim 2019; Rodriguez-Ramos 2018; Schwab 2017; Sebastian et al. 2017; Westerman et al. 2011; Williams and

Schallmo 2018). Still, clusters based on Briggs's (2019) more granular annual tech-trends research were found to best summarize what we need for the purpose of this book. Admittedly very subjectively, I reordered them in decreasing order of relevance, at least from what I know now.

Table 4.1 structures all later-explained technology based on their assumed maturity and their potential impact at a larger scale.

TABLE 4.1   **Digital technologies overview.**

	Lower Maturity	Higher Maturity
Higher impact at scale	IV. Analytics technology VI. Cyber-technology	I. Cloud technology II. Digital experience technology
Lower impact at scale	VII. Digital reality technology VIII. Blockchain technology VIII. Longtail technologies	III. Intelligent automation technology

Cloud Technology

The most relevant enterprise technology and supply-side technology catalyst for digital transformation during the last decade was *cloud technology* (Briggs 2019). It is clearly among of the biggest shifts in the history of IT since the introduction of the internet.

Originally pioneered by VMWare to lower the cost of deploying technology by using central capacity and resources (Rodriguez-Ramos 2018), cloud technology has moved away from a mere infrastructure cost lever to a potential catalyst for large-scale delivery and business model innovations and transformations. It now goes beyond data-center modernization and is, in terms of technology, dominated by well-known innovation powerhouses like Amazon AWS, Google, Microsoft, IBM, Alibaba, and Tencent.

Driven by the multiple promises of the technology, many businesses already have moved or are in the process of migrating most of their workloads (applications and data) into the cloud in its numerous forms (inhouse, co-location, hybrid, public, . . .). "From infrastructure-as-a-service (IaaS) capabilities, to platforms-as-a-service (PaaS), to now a growing ecosystem of vendors continue to methodically automate ever-higher order processes to create industry-optimized platforms" (Buchholz and Briggs 2022). The workloads affected cover the whole value chain from customer-driven front-ends to back-end oriented finance systems. However, whenever you get the opportunity to look deeper into the underlying program designs, you will find most of these transformations to be technically driven under high time pressure with a limited or only retroactively constructed link to the overall business strategy. In addition, in a lot of companies, even if the business-strategy link does exist, it is then not correctly translated into IT-strategy implications. Think about your company. Did you, like many others, start rushing into the cloud some time ago? Did you fail to synchronize your multitude of cloud migration projects and end up with numerous cloud suppliers? Did this problem introduce consequences regarding processes, security, and governance? Do now neither internal nor (even worse) external customers feel any of the originally targeted benefits? I would not be surprised. The cloud is hardly ever used to its full potential. No wonder. From the outset, the business levers for harvesting this very potential were never clearly defined. In many cases, programs thus end up as a legacy infrastructure replacement exercise, not at all leveraging what they could do beyond infrastructure cost savings. Even more well-designed programs are often only beneficial in comparison to a (by the way, very hard to derive) cost baseline. They use operational key performance indicators (KPIs) (e.g., from DevOps and other tools) and, in the best cases, some additional indirect KPIs to better understand the cloud's platform effect to later exercise the real option on the stacked-on business value.

I almost never see cloud transformations designed to target and measure what *really* matters to the business (as a strategic business objective): the improvements for end-customer and internal customer journeys that cloud transformations can bring (e.g., measured improved customer satisfaction scores like net promoter score (NPS) or customer satisfaction index (CSI)) and all the corresponding positive strategic and financial implications beyond cost savings).

And as if all of this would not be frustrating and challenging enough, you also have to cope with what has recently been brought to the attention of practitioners as the cloud paradox (Wang and Casado 2021). It suggests simply and painfully that it is "crazy to not start in the cloud [and] crazy if you stay [only] on it." Why? Because the splendors of all the payday accelerators described in the following sections may not remain valid if you are not growing at the same pace as your exploding cloud costs. At some point, this can eat into your margins in a way you never expected. Ironic, or not? Especially since expected cost savings were and are the main reason why businesses choose to go to the cloud in the first place. Today, the admittedly counterintuitive idea that at some point there might be a need of "repatriating" some workloads from the cloud to own infrastructure or execute a cloud-to-cloud migration to another vendor is a major task. This is evermore true once you make all your infrastructure experts redundant, because you were so sure they would no longer be needed.

Before any of this comes into play, however, a major source of concern remains finding solutions for the issue of massive technical and technology debt (Magnusson and Bygstad 2014). Core IT needs to shed its siloed past and develop a more cross-functional approach in which IT and business act as one bimodal (Haffke, Kalgovas, and Benlian 2017) or dual-speed (Westerman, Bonnet, and McAfee 2014) integrated team. Given the complexity of their technical legacy environments, organizations aim to "show increased capability to reinvigorate their legacy core by exposing micro services to their technologists and the business" (Briggs 2019, p. 10).

So, when you are embarking on or continuing your journey to take workloads into the cloud you must make sure that the long-term business rationale and approach behind doing so is very clear. Do you proceed based on a reasonable and resilient value case step by step—for example "Lift'n Shift"—by partial 1:1, porting of workloads and successive migration (rehost, replatform, refactor, recode, repurchase, retire) or go for a big bang? To decide, you must think about a large set of complex accelerators and decelerators, which will influence the timing and magnitude of your digital transformation payday.

Cloud Technology Payday Accelerators   Once you have managed to migrate a relevant share of your external customer-facing and internal workloads, or of enabling infrastructure to support a superior performance of your front-ends into the cloud, your new cloud solutions can ideally be integrated with your digital experience technologies (explained later) and better support end-to-end customer journeys. This indirectly can increase related revenue and efficiency drivers. These you can model, target, and measure in your cloud migration business case. They include improved customer satisfaction scores, increased conversion and retention rates due to increased ease of use in front-ends, and increased automation with less manual process steps in front-end/back-end integration. In addition, as I will explain in the Analytics Technology section of this chapter, the ability to combine data from various sources in one place (in the cloud) allows a wide range of additional benefits and new business models.

This business case upside has the potential to be boosted even more if you simultaneously aim to have more agility and faster time to market in order to react to upcoming market needs. Or, if you can outpace competition as a first mover with new propositions by carefully analyzing your data trove and applying analytics and pattern recognition. You can plan these impacts and measure them as incremental revenues from first-mover product-and-services launches, profit improvements due to faster and segmented lifetime value-driven service portfolio and pricing adjustment, or as the indirect benefits of faster customer journey improvements and process-driven cost savings.

At the same time, your cloud solution can provide greater scalability and close-to-instant availability. This generates better capacity utilization versus cost ratios, which you can model in your business case as enablers of revenue growth cases or as mitigation to demand forecasting uncertainty/fault cost in case of massive volume peaks or surges.

Furthermore, you can aim to reduce your risks of service disruption, which can usually be extrapolated and estimated as cost avoidance or opportunity cost from related historical developments. Or you can forecast in different disruption-threat scenarios, ideally closely linked with unfortunately high-cost cyber-technology considerations. These will be elaborated in much more detail later. The fact that (cyber) security is now under the responsibility of the cloud providers can provide synergies. This can, for example, be a shared cybersecurity team or immediate security updates in the cloud without the need of lengthy upgrade rollouts.

Such synergies can offer you cost advantages like as the best available services and features developed centrally by the vendors, and increased overall interoperability and flexibility. Obviously, depending on the cloud provider, this also means that you will lose autonomy and become a consumer of services instead of the driver of development. At the same time, you can substantially increase process efficiency in operations via new approaches for development cost reduction, which can be modeled in your business case when planning the corresponding cost buckets.

Finally, moving to the cloud can free you up from previous capex constraints, by moving the majority of spending into the opex bucket, which, as mentioned before, can also be a two-sided sword—on the one hand, if the margin impact affects the view of investors on your valuation negatively, on the other hand, if headcount reductions will be feasible due to the fact that experts on your legacy systems and infrastructure topics will no longer be needed in-house.

Cloud Technology Payday Decelerators Unfortunately, every cloud transformation also faces a wide range of payback decelerators, which I often see underrepresented in business cases. These decelerators start with (multi-) cloud strategy design cost, a crucial investment, which is often overlooked, leading to a purely technology-driven cloud transformation program, followed by the often substantial cost (if done right, because usually external neutral expertise is required over a long selection time period) for selecting the best fitting vendor or vendor portfolio in the light of the previously developed strategy, license, and usage cost and the often difficult-to-plan implementation and rollout cost from vendors, staff augmentation, subject-matter experts, and consultants. As in any program, you should plan with a substantial buffer over and beyond what you think will be required.

This might technically take you to the cloud, but unfortunately required spending does not end here to generate the business impact you should be aiming for. Your payday will be further slowed down by softer sounding but often success-critical hiring and training cost to replace the expensive externals you had to onboard in the early parts of your journey with very scarce and highly paid full-time cloud expert staff; retention packages for already skilled users, developers, and architects you might have on board; and change management cost for the broader organization to actually appreciate and use the new technical, process, and innovation capabilities in line with your overall strategic goals. (All these workforce-related topics will be further elaborated later in this chapter.)

Ongoing license, operations, and maintenance costs will obviously have to play another important role in your payback considerations, especially as the previously discussed value implications of impeding margin pressure after capex to opex shift could negatively influence your market capitalization from an external perspective. You could be stuck in a vendor lock-in, potentially with substantial minimum-volume commitments, without a proper exit and migration strategy to divorce from the relationship at a later stage if economics require you to do so. Cloud transformation is no one-time effort but, rather, an ongoing process with permanent adaptations (new releases, new skills, and regulatory requirements).

Last, but often substantially underestimated, are the security costs any cloud transformation will add to your baseline, whether you like it or not. Buckets to include in your business case are data security validation efforts, regulation-driven adjustments (e.g., from GDPR), and other contractual obligations.

Table 4.2 provides a summary that clusters some key accelerators and decelerators for cloud technology depending on their typical tangibility.

Digital Experience Technology   *Digital experience technology* has become an umbrella term for architectures integrating a complex range of marketing, sales, CRM, and service technologies and platforms. Historically built for customer-facing digital marketing and e-commerce solutions, digital technology is now evolving more and more. It is developed with a strong focus on experience, to enhance "all the ways organizations, customers, employees and constituents engage and carry out transactions within digital environments" (Briggs 2019, p. 7). Digital experience technology works as a second supply-side catalyst and changes the way firms act to create value for their customers. It is now envisioned to cover the entire enterprise either by one technology platform or by a tightly integrated portfolio of best-of-breed solutions, often combined with paradigms like microservices and headless approaches.

Large software vendors like Salesforce, SAP, Workday, ServiceNow, just to name a few, use this widespread paradigm shift to customer centricity to push new solutions into the market, advertising to offer new means to lower the barrier to digital transformation across all segments and channels.

Next to the big vendors just mentioned, an almost infinite number of technology companies in all different shapes, forms, and sizes aims to achieve two very simple things from a business perspective: First, to get the best out of a more intimate knowledge of the customer and enable acting based on value-driven considerations at the customer interface, shared across all your channels and consistent in communication and customer lifetime view. Second, to ensure a seamless processing throughout the enterprise of whatever comes out of these targeted customer interactions. On the one hand, this requires a portfolio of solutions and a piece of software that combines data from multiple tools to create a single centralized customer database containing data on all

TABLE 4.2 Cloud technology payday drivers.

Drivers	Less Tangible	More Tangible
Accelerators	• Indirect support of customer satisfaction, conversion-and-retention rates due to increased ease of use in front-ends and increased end-to-end automation • Fundament for analytics innovation plus new use cases and revenue streams • More agility and faster time to market • Reduced risks of service disruption, (cyber) security synergies on the central vendor platforms	• Process automation driven cost savings • Development cost reduction (DevOps) • Operations effort reduction (but Capex to Opex shift) • Expert personnel cost savings (infrastructure, legacy system developers and so on)
Decelerators	• Multicloud strategy design cost • Vendor or vendor portfolio selection cost • Hiring and training cost • Retention packages for already skilled users, developers, and architects • Change management cost • Cost for ongoing adaptations (releases, skills, regulatory requirements) • Security cost for data security validation efforts, regulation adjustments (for example, from GDPR) • Vendor-exit strategy and migration cost • Risk buffer	• Ongoing license, usage, operations, and maintenance cost

touchpoints and interactions with product. On the other hand, a more systematic approach to act in a truly customer-centric way becomes key. No longer is the process transaction from an internal perspective the focus but, rather, a full suite of end-to-end technologies from the website to online shop, to the middleware layer, and all the way to the integration with your back-end systems.

In any case, any effort to leverage digital experience technology must make sure that your long-term business requirements are very clear and in line with your business strategy. At the same time, you need to establish a mindset that constant small changes and an adaptive approach are as important as working within the overall strategic vision.

I have often seen business requirements either just statically replicating the complex legacy status quo or, in somewhat better cases, aiming for improvements but lacking the answers on the true business questions that matter. Implementing digital experience technology effectively is a multiyear journey. Will the results put the company in a differentiating position after launch, or does the program end up becoming an exercise that does not differentiate from the competition, because everyone is doing the same thing?

Unfortunately, this means that digital experience technology does not only provide many nice-sounding accelerators but will, in many cases, also imply high-cost decelerators, which will influence the timing and magnitude of your digital transformation payday.

Digital Experience Technology Payday Accelerators   Once you have all your digital experience platforms up and running, ideally as specified in an iterative process in which business and technology teams work closely together, the benefits of digital experience technologies can be manifold—at least if you manage to achieve a relevant and never-resting capability boost versus your legacy systems. Then you can expect incremental revenues from many sources of new customers and within your existing customer base. These could take the form of:

- Higher conversion rates from your campaigns
- Lower acquisition costs
- Incremental revenue uplifts from better matching customer needs and (next best) offers when upselling and cross-selling
- More successful identification and retention of valuable at-risk customers
- Phase-out or migration of lower- or negative-profit customers to other services

However, beware. These are part of the flashy hypothesis, which every 360-degree vendor postulates. You need proof from experience or, even better, from properly executed pilots, before you can rely on any numbers for your business case.

Digital Experience Technology Payday Decelerators   Digital transformation failure rates have one very important source of origin. In real life almost every digital experience implementation is fighting expected and unexpected payback decelerators, which often come as a surprise for management when they materialize, even though they should have been foreseeable. First and foremost, what you should factor in are the often-unaccounted-for cost for a preproject and a feasibility study to be surer of what you are doing prior to starting into the larger exercise. Iterative, lean, approaches must be constantly adjusted based on feedback and change. They require you to start early, fail, do it again, and correct your trajectory to do it right in the end. Design exercises that are too big up-front only lead to inertia and projects not being started. These are all business design costs in the wider sense. Often overlooked are also the substantial cost (if done right, as usually external neutral expertise is required over a long selection period) for selecting the best fitting vendor or vendor portfolio in the light of the previously developed strategy. More transparent and plannable are license and usage costs, less so the often difficult-to-plan implementation and rollout costs from vendors, staff augmentation, subject-matter experts, and consultants. I have seen many cases in which the initial happiness of selecting the cheapest solution turned out to be a rude awakening when change requests blew the budget to a magnitude no one expected. Very simply, if you buy cheap, you often buy twice, so a substantial risk buffer is advisable in any case. All this might make your systems go live one day, but required investments do not end here. Also to be considered are:

- Success-critical hiring and training
- Onboarding cost to replace the inevitable externals you had to onboard in the early parts of your journey with very scarce and highly paid full-time expert staff (either for implementation or for backfilling, as you still have a business as usual to run in parallel)
- Retention packages for skilled users, developers, and architects you might already have on board
- Change management cost for the broader organization to actually appreciate and use the new technical, process, and innovation capabilities in line with your overall strategic goals

All these workforce-related topics will be further elaborated in separate workforce catalyst sections of this chapter.

Ongoing operations and maintenance cost will obviously have to play another important role in your payback considerations.

Last, but often substantially underestimated, are security costs. Buckets to include in your business case are data security validation efforts and regulation-driven adjustments (e.g., from GDPR), which I will explain in more detail in a later section.

Table 4.3 clusters some key accelerators and decelerators for digital experience technology, depending on their typical tangibility.

Intelligent Automation Technology   The third increasingly relevant supply-side technology catalyst is more a group of technologies than a single innovation. It includes "technologies such as machine learning (ML), neural networks, robotic process automation (RPA), bots, natural language processing (NLP), and the broader domain of . . . (AI) . . . and can help make

TABLE 4.3 Digital experience technology payday drivers.

Drivers	Less Tangible	More Tangible
Accelerators	• More effective upselling and cross-selling of new products and services	• Higher conversion rates and lower acquisition cost from proven products and services • Better retention of valuable at-risk customers • More targeted phase out or value-based migration of lower- or negative-profit customers
Decelerators	• Preproject, feasibility study cost • Vendor or vendor-portfolio selection cost • Hiring, training, and onboarding cost • Change management cost • Retention package cost for skilled users, developers, and architects • Technical implementation and rollout costs • Cost for staff augmentation, subject-matter experts, consultants for implementation support and backfilling • Cyber- and data security cost, like data security validation efforts and regulation driven adjustments • Risk buffer	• License and usage cost • Ongoing operations and maintenance cost

sense of ever-growing data, handling both, the volume and complexity that human minds and traditional analysis techniques cannot fathom" (Briggs 2019, p. 9).

These intelligent automation technologies in practice already support manifold use cases ranging from automated contract and claims handling in finance and procurement, to cognitive solutions like virtual chat- and voice-bots in customer service and HR/recruiting, to Internet-of-Things (IoT)-data-supported predictive maintenance, to causal machine learning in customer experience impact analysis, to AI-driven large-volume anomality recognition in supply chain and cybersecurity management, just to give a few examples.

The business reasons for implementing such intelligent automation solutions are usually quite intuitive. However, real-life examples have consistently shown that overselling the financial impact of a first pilot (after an initial proof of concept) and the ability to scale up across the enterprise at high speed is usually not the best idea. Many projects I have seen struggle to achieve what they originally promised financially and often end up being associated to "softer" benefits at delayed timelines and with limited scale impacts.

Intelligent Automation Technology Payday Accelerators Rarely have I seen the intelligent automation solutions lead to full headcount reductions as the naivest and, from a worker's perspective, most feared benefit of adding "intelligent robots" to the workforce. Instead, even though there are exceptions where three-digit number of headcounts have been reduced, in most cases the current state of available tools (especially RPA) rather allows part of the workload of personnel being given back to the business, not the full-time equivalent headcount. This is achieved by offloading staff from cumbersome, inefficient, and, in most cases, boring and not very fulfilling tasks. Nevertheless, the impacts should not be underestimated: Companies with very high growth trajectories can shoulder the implied additional workload without significantly increasing their workforce. This is driven by the indirect impacts automations can have on compliance or risk pass/fail data, net promoter scores, process speed, first solution rates, and many more. Still, for real transformational personnel cost impact, processes cannot just be painted over with new intelligent automation tools but need to be completely redesigned as part of the overall digital transformation effort. Intelligent automation solutions can serve as an additional booster of automation, as a positive "modern" driver to start previously stuck process-optimization discussions or as a bridging technology until the full-scale process redesign becomes available.

Intelligent Automation Technology Payday Decelerators On the other hand, cognitive technology driven payback decelerators are much more concrete and immediate. You must plan cost for RPA, cognitive/AI software (licenses, support cost), and related bolt-on solutions (e.g., OCR), the necessary hardware or cloud hosting cost, plus plan to invest substantial time and money into hiring, training, learning on the job, change management, and ramp-up inefficiencies until the integration with other work steps is practiced well enough. Under the assumption that in the early days you do not have scalable cognitive/AI expertise in-house, you will also face highly relevant cost for required external expertise: to raise awareness for the success-critical process of scouting and mining (potentially by different software with additional cost), to find the right pilot processes with buy-in and scalability potential, process detail analysis and documentation, platform

TABLE 4.4   **Intelligent automation technology payday drivers.**

Drivers	Less Tangible	More Tangible
Accelerators	• Impacts automations can have on compliance or risk pass/fail data, net promoter scores, speed, first solution rates and many more	• Personnel capacity given back to the business • Some full-time equivalent headcount savings
Decelerators	• RPA/cognitive/AI software (licenses, support cost) and bolt-ons (OCR, . . .) • Hardware and/or cloud hosting cost • Hiring, training, learning on the job, change management cost • Ramp-up inefficiencies • Cost for external expertise, to raise awareness, process scouting and mining, detail analysis and documentation	• Cost for implementation and configuration, go-live/hyper care and ongoing maintenance & support

implementation and configuration, go-live/hyper care, and ongoing maintenance and support.

Table 4.4 clusters some key accelerators and decelerators for intelligent automation technology depending on their typical tangibility.

Analytics Technology   Leveraging the massively growing supply of continuously generated data (Rogers 2016) is the obvious next step for value creation, once the previously discussed cloud technology with superior computing capabilities (Brynjolfsson and McAfee 2014), centralized storage, and the possibility to analyze information at scale overcomes previous economic barriers (Rodriguez-Ramos 2018) of data storage and aggregation. As soon as your digital experience platforms provide you with and give access to the data you need, you can leverage "big data," structured and dark, inside and outside firm boundaries, analytics engines, algorithms, and supporting infrastructure, ideally combined with cognitive technologies (described later), to serve as a catalyst for firms to predict outcomes and recommend derived value-creating actions at scale.

Analytics Technology Payday Accelerators   Unfortunately, until recently, "most analytics efforts have struggled to deliver on the simplest version of that potential: the rearview mirror describing what has already happened—or, for the advanced few, presenting real-time views into what is currently happening" (Briggs 2019, p. 7). Still, many analytics experts take it for granted that analytics should be a critical future cornerstone for any successful digital transformation. For them it seems obvious that a substantially growing annual analytics budget should be approved without any doubts and they are then surprised when they face resistance because the payday for analytics technology is not that easy to explain and demonstrate to their business. That does not mean that, in theory, the benefits of a successful analytics platform and team are not clear: With advanced analytics, you can better target which customers to acquire and keep for optimum value and how to best cross- and up-sell to them. Furthermore, instead of using only basic measurements for decision-making you can improve the effectiveness of your marketing attribution and media mix modeling and thus optimize your advertising spend. Analytics also allows you to get a better understanding of correlations and, in the best case, something close to causality between planned action and outcomes. On top of that, analytics also leverages data to do things faster. In combination with cognitive/AI catalysts (as explained later in this chapter), you can act much quicker without the previous manual analysis. The payday contribution of speed can be measured and demonstrated, for example, as a short-term resource time and expense reduction for developing and deploying reports, and the longer-term revenue gains and expense reductions from quicker performance analytics being used to streamline business process and identify opportunities for growth.

Analytics Technology Payday Decelerators   On the other side of the equation, analytics payback decelerators are not to be forgotten. You must plan software cost (licenses, support cost), the necessary hardware or cloud hosting cost, and expect to invest substantial time and money into hiring, training, learning on the job, change management, and ramp-up inefficiencies until the integration with other work steps is practiced well enough. Under the assumption that in the early days you do not have all the analytics expertise in your workforce, you will also face cost for external expertise: to raise awareness, to find the right use cases with buy-in and scalability potential, process detail analysis and documentation, platform implementation and configuration, go-live/hyper care, and ongoing maintenance and support.

Last, you should not forget implied security cost. This includes data security validation efforts and regulation-driven adjustments (e.g., from GDPR).

Table 4.5 clusters some key accelerators and decelerators for analytics technology depending on their typical tangibility.

Cyber- and (Data) Security Technology   All technology capability catalysts come at a substantial risk. Hostile security attacks plus regulatory data security and privacy boundaries from numerous stakeholders start to play an increasing role in any digital transformation: "With cybercrime

TABLE 4.5   **Analytics technology payday drivers.**

Drivers	Less Tangible	More Tangible
Accelerators	• Better target which customers to acquire and keep for optimum value and how to best cross- and upsell to them. • Improve the effectiveness of your marketing attribution and media mix modeling • Gain better understating of correlations and in the best case something close to causality between planned action and outcomes	• Leverage data to do things faster (for example, reports)
Decelerators	• Hiring, training, learning on the job, change management cost • Ramp-up inefficiencies • Cost for external expertise, to raise awareness, use case selection, detail analysis and documentation • Security cost • Risk buffer	• Analytics software (licenses, support cost) • Cost for implementation and configuration, go-live/hyper care and ongoing maintenance and support • Hardware and/or cloud hosting cost

estimated to cost US$ 6 trillion annually by the end of this year, cloud migration raises the cybersecurity stakes" (Golden and Kunchala 2021; Morgan 2020). As omnipresent threats or at least implicit or explicit business limiters and enablers, they can become catalysts by themselves. "Companies are pushing the boundaries of the security function and shaping their risk appetite before development begins. Going forward, cyber will undergird every component of the macro platform, and will be integrated into . . . all aspects of an organization's digital . . . agenda" (Briggs 2019, p. 11). In other words: Without a proper (cyber-) security transformation element integrated right from the start, no digital transformation effort will survive long enough to create any value.

So, while risk-mitigation measures against attackers and for enhanced customer data security and privacy protection cannot be business objectives in isolation, they still become highly relevant as a make-or-break factor. In the age of cyber-attacks, data thefts and prominent regulatory fines for data protection infringements, the assurance of integrity, confidentiality, and availability of services are no longer just an IT issue but a top management priority for any (digital) business transformation program.

This is even more true when you can expect the number of attacks and potential regulatory data security and privacy violations to grow exponentially with the number of cloud platforms and workloads planned to be migrated. In such a complex world outside of your direct influence, estimating related digital transformation payday accelerators and decelerators becomes a critical but very challenging task.

Cyber- (and Data) Security Technology Payday Accelerators The general problem of estimating cybersecurity and data protection technologies as a digital transformation payday accelerator is that they are abstract. They are designed to maintain customer trust and secure ownership of one of the most crucial assets of your organization—data. As such, next to enabling new business models, access to new markets, and strengthening of value propositions, they are first and foremost about loss prevention. When you invest in security, you aim to reduce risks threatening your assets. The return on security investment is calculated by estimating how much loss you avoided thanks to your investment. This requires approximating how high these damages in various forms could have been if nothing had been done to mitigate them. Unfortunately, a proper estimation requires replicable processes and correct data to be captured. At many companies, I know these do not exist at the level of quality you would expect them to be. The often-necessary ad-hoc gathering of information usually diminishes buy-in of recommendations before even getting started. And sadly, even benchmarks do not help much here, because every enterprise is different.

There is certainly no lack of concepts (ENISA 2012) to still get a good estimate. They share the intuitive idea to approximate the damage a certain event would imply, the single loss expectancy (SLE) multiplied by the likelihood of its occurrence (the annualized risk of occurrence, or ARO), giving you the annual loss expectancy (ALE). Whatever your measures, subtract from this ALE the benefits you are looking for. Obviously, these benefits are very different in character compared to the digital transformation payday accelerators discussed earlier. The first key question is: What are the damages that are being avoided? They range from the implications of non-adherence to regulatory requirements (to protect licenses, for example, in financial services), to data privacy violation linked penalties (GDPR), to related personal liabilities of board members. Also, the damages of a loss of trust must be considered, regardless of whether it comes from your customer base or from your ecosystem partners (if business-critical information is visible to competitors) or if private data (for example, on financial status) is becoming public. Not to forget real damages (data theft and blackmailing, critical system outages, and many more). In any case, the annual rate of occurrence is hard to estimate, and the resulting numbers can vary highly from one environment to another.

These approximations are often biased by our perception of the risk. The accuracy of statistical data used in the calculation of return on sustainability investments (ROSI) is therefore essential. However, actuarial data on security incidents are hard to find as companies are often reluctant to capture data on their security incidents.

Cyber- (and Data) Security Technology Payday Decelerators   The cost of the required solutions and services is easier to predict provided all payday decelerators are considered. They can be best structured by asset classes:

1. First, one must consider the cost for dedicated hardware whose prime purpose is IT security related, including firewalls, security gateways, security appliances, security toolset platforms, and ID tokens.

2. Second, software cost like annual license and maintenance as well as costs associated with new purchases and upgrades for all software dedicated to operate or manage the security systems applications for each category of security expenditure are of relevance.

3. Third, and not to be forgotten, facility and other costs such as hosting/ facilities/occupancy for space dedicated to in-scope security functions and personnel, such as the apportioned annual costs of hosting security-related devices, storage arrays, and appliances in the data center, including power/heat management and raised floor. It would also include the annual cost of any consumables related to the security activities.

4. Fourth, outsourcing fees for third-party or outsourcing contracts primarily comprising services for managing or monitoring security devices, systems, or processes where the services are provided on-site should be an integral part of the business case.

5. Fifth, the cost of managed service providers (MSPs)/cloud as remote subscription-based monitoring and/or management of security devices such as firewalls, intrusion detection, and prevention functions via customer-premises-based or network-based devices can become a key digital transformation payday decelerator. It also includes remotely delivered specialist-managed security services (e.g., threat intelligence, security information and event management/security operations (SIEM/ SOC) center, distributed denial-of-service attack (DDoS), etc.) and cloud-based security services such as identity-as-a-service (IDaaS).

6. Sixth, consulting services that help companies analyze and improve the efficacy of business operations and technology strategies.

7. Seventh, IT security personnel whose roles and duties are primarily focused on information security activities can become a key cost block. This includes all full-time, part-time, and temporary full-time equivalent resources (FTEs). These personnel provide support across the security functions.

TABLE 4.6 Cyber- (and data) security technology payday drivers.

Drivers	Less Tangible	More Tangible
Accelerators	• Prevented post-incident loss of trust from your customer base • Secure enablement of new business models	• Avoided fines from non-adherence to regulations, data privacy violations (GDPR), related personal liabilities of board members
Decelerators	• Security consulting services • Risk buffer	• Dedicated hardware cost (firewalls, gateways, security appliances, toolset platforms and ID tokens) • Software cost (purchase, annual license, and maintenance) • Annual costs of hosting security-related devices, storage arrays and appliances in the data center • Outsourcing fees (for managing or monitoring security devices, systems, or processes) • Cost of managed service providers (MSPs) • IT security personnel cost

Table 4.6 clusters some key accelerators and decelerators for cyber-technology depending on their typical tangibility.

Digital Reality Technology   The sixth supply-side technology catalyst for digital transformation is more a cluster of technologies than a technology itself. Way before the hype of the Metaverse became the talk of the town, immersive or even implantable (Schwab 2017) technologies like augmented reality (AR), virtual reality (VR), mixed reality (MR), and Internet of Things (IoT) based on a substantial evolution of mobile and fixed network infrastructures plus capacity were foreseen "redefining how humans interact with data, technology, and each other" (Briggs 2019, p. 8). This goes hand in hand with conversational interfaces, computer vision (Schwab 2017) and other recent innovations like the "metaverse" (Foutty and Bechtel 2022).

All these technologies are meant to ultimately replace the established human-technology interface and can be used in marketing and sales, event/meeting management, in field services, and in training and immersive visualization of products and locations. No matter what the use case is, the abstract benefit claim is always the same—these technologies will generate new sources of revenue, increase productivity, or improve safety. Nevertheless, when you start to look deeper, in most use cases the underlying digital transformation payday accelerators and decelerators are very problematic to pin down more concretely.

Digital Reality Payday Accelerators Unfortunately, scaling up digital reality solutions in a meaningful way first requires a sufficiently large hardware user base in your organization. A few niche pilots will not move the needle in any way you can measure. Not only do you need many users to be able to one day find any meaningful benefits in your bottom line, but you also need them to use the devices for their respective use case very frequently to pay back the required hardware investments. This is certainly true for the most obvious and communicated training use cases (savings in travel and location cost, steeper learning curve, and so on) but even more for use cases in field services (reduced training cost, remote direct access to specific subject-matter experts, and so on), not to mention the required customer take-up rates required for any selling and marketing efforts in digital reality to make any sense from a reach-and-impact perspective.

Digital Reality Payday Decelerators Next to the cost of sourcing (buying or as a service) the necessary hardware and software platforms to achieve sufficient penetration in your workforce for the selected use cases, creating digital reality scenarios at scale requires a specific up-front investment in terms of content creation. This can initially be external (agencies, consultants, subject-matter experts), but if a certain scale must be reached, it requires also building up internal expert resources, which are forecasted to be high-cost, even today.

Decelerators also include strategy design, followed by the substantial cost for selecting the best-fitting vendor or vendor portfolio in the light of the previously developed strategy, license, and usage cost and the often difficult to plan implementation and rollout costs from vendors, staff augmentation, subject-matter experts, and consultants. As in any innovative program, you should plan with a substantial buffer.

Your payday will be further slowed by softer sounding but often success-critical hiring and training cost to replace the expensive externals with expert staff; retention packages for already skilled users, developers, and architects you might have on board; and change management cost for the broader organization to actually appreciate and use the new technical, process, and innovation capabilities in line with your overall strategic goals. Ongoing license, operations, and maintenance cost will obviously have to play another important role in your payback considerations.

Last are the implied security costs, and today we have not even begun to comprehensively assess them. Buckets to include in your business case

TABLE 4.7 Digital reality technology payday drivers.

Drivers	Less Tangible	More Tangible
Accelerators	• Steeper learning curve • Remote access to specific subject-matter experts • New immersive virtual markets with marketing and sales potential	• Savings in travel and location cost for trainings, meetings
Decelerators	• Strategy design cost • Use case selection cost • Vendor or vendor portfolio selection cost • Content production cost • Hiring and training cost • Change management cost • Cost for ongoing adaptations (releases, skills, regulatory requirements) • Security cost • Risk buffer	• Hardware and platform cost • Ongoing license, usage, operations, and maintenance cost • External advisory cost

are data security validation efforts and regulation-driven adjustments (e.g. from GDPR).

For easier handling in your daily business, Table 4.7 clusters some key accelerators and decelerators for digital reality technology depending on their typical tangibility.

Blockchain/Distributed Ledger Technology   Originally mostly known as the underlying technology of cryptocurrencies (Rodriguez-Ramos 2018), blockchain and other (still) less prominent distributed ledger technologies (DLT) as supply-side technology catalysts or shifts (Schwab 2017), begin to transform the crucial matter of trust in more and more business interactions and contractual transactions beyond trendy media headlines (Walker and Hansen 2021). They serve as "a profoundly disruptive technology that transforms not only business but the way humans transact and engage . . . with technical hurdles and policy limitations [now] being resolved, we will likely see breakthroughs in gateways, integration layers, and common standards in the next few years" (Briggs 2019, p. 8). Simply put, blockchain implementations can establish a more secure shared information vehicle, a single source of truth as a resource that is trusted by multiple parties. It can thus replace siloed data storages with biased owners and is, by design, much easier to protect against external security threats.

On the blockchain phenomenon, you will find two different views among strategists and technologists. They either see it as an overhyped longtail phenomenon or, and this is the growing faction and includes me, as a potential longer-term solution to business challenges in highly interconnected ecosystems (for example, supply chains), that is across multiple entities; as a new answer when other technologies have never succeeded. Unfortunately, both camps mostly delve on the excitement on, or criticism of the same blockchain-related use cases. And there are many, which are gaining traction: self-sovereign data and digital personal identity, trusted data-sharing among third parties, grant funding, intercompany accounting, supply chain transparency, customer and fan engagement, creator monetization (Buchholz and Briggs 2022).

They still often miss the core of what business leaders really want to know: When, if ever, is the payback on blockchain implementation happening at scale? Not just for small pilots, but within a timeframe that matters for the business overall and not just for niche business problems. To make things even more challenging, blockchain projects share many commonalities with other technology implementations but add one unusual complexity. They are meant to serve an ecosystem, where, as in the smart-contracts business model, not every participant is under your direct influence. Therefore, forecasting the potential payday from blockchain becomes a serious challenge. Digital transformation payday accelerators will often be intangible (like uncovering a serious supply chain issue in real-time) and long-term, while decelerators are much less so.

Blockchain/Distributed Ledger Technology Payday Accelerators

The primary abstract benefit of blockchain applications in an ecosystem will always be trust. Once the technology has replaced the traditional human-relationship-based notion of trust among the partners with a technical algorithm, your business can expect revenue increases from transactions otherwise not possible. Hand in hand with trust comes the advantage that blockchain creates unalterable records with strong encryption and is stored across a network of computers, making it very challenging to attack. It goes without saying that both these increases are very difficult to translate into concrete revenue increases. Easier to estimate are the implications on the costs for organizations by creating process efficiencies and easing reporting and auditing processes. Blockchain also helps businesses cut costs by removing third-party providers, and by automating processes in transactions, blockchain can operate transactions significantly faster in a much more traceable manner, which should also be recorded as a cost benefit on the accelerator side of blockchain/distributed ledger technology (DLT) business cases.

Blockchain/Distributed Ledger Technology Payday Decelerators
Unfortunately, the described abstract benefits come at a price. Every blockchain implementation is also producing a range of much more concrete payback decelerators. These decelerators start with the blockchain use case selection and partner search and contracting cost, both crucial investments to get right for later scalability in the ecosystem (or by joining a consortium) and continue with the cost for selecting the best-fitting vendor or vendor portfolio in the light of the previously selected applications. Most of the time, you will only have small pockets (if any) of expertise in-house and will have to consider the expense of bringing subject-matter experts on board, either via traditional consulting projects or by hiring (temporary) expert staff to start building up the required capabilities. As with all other technologies discussed in this chapter, this might take you technically ahead, but, unfortunately, required spending does not end here to generate the scalability required for the blockchain to make sense. Your payday will be further slowed down by often success-critical hiring and training cost to replace the expensive externals you had to onboard in the early parts of your journey with very scarce and highly paid full-time blockchain expert staff, retention packages for already skilled users, developers and architects you might have, and change-management cost for the broader organization to actually appreciate and use the new technical, process, and innovation capabilities in line with your overall strategic goals.

Ongoing license (if no open-source stack is used), hosting, operations and maintenance cost will obviously have to play another important role in your payback considerations. Blockchain is no one-time effort but an ongoing process with potential adaptations (new releases, new skills, regulatory requirements). While the security cost benefits have already been discussed on the accelerator side, blockchain implementations at scale also might, like any other IT implementation, carry security cost implications (which have not yet been fully reflected by regulators but for sure will be, not to mention the potential dangers of new technologies like quantum computing) with them, which should not be underestimated. Buckets for your business case are data security validation efforts, and regulation-driven adjustments (e.g., from GDPR).

Table 4.8 clusters some key accelerators and decelerators for blockchain technology depending on their typical tangibility.

Longtail Technologies   As always in the digital transformation age, the attitude of looking for the next big thing and the opportunities underlying these trends becomes crucial. That leads to the question, What's next? Next to more long-term visions (sources), there are many resources you can use to build your opinion for the midterm, proven and unproven. For this I usually rely on the technology trends work (Buchholz and Briggs 2022), which my research colleagues have built over the last years. Several big themes will become relevant over the next years. They are covered in the following subsections.

TABLE 4.8   Blockchain technology payday drivers.

Drivers	Less Tangible	More Tangible
Accelerators	• Revenue upsides from new transactions based on algorithm-based trust is established among the participants • Increased security due to unalterable records with strong encryption and storage across a network of computers	• Cost savings due to process efficiencies, ease of reporting, traceability, and simplification of process auditing • Cost savings due to removal of third-party providers
Decelerators	• Cost of blockchain use case selection and partner search and contracting cost • Hiring and training cost to replace expensive externals • Change management cost • Security cost like any other technology	• Cost of bringing subject-matter experts on board, either via traditional consulting projects or by hiring (temporary) expert staff • Ongoing license (if not open-source), hosting, operations, and maintenance cost

Next-Generation Infrastructure Technology   You could argue that fiber rollouts, 5G/6G networks, IoT, and smart infrastructure are already here today and therefore should not be put as a future prospect here. Yes, they are. But believe me as a telecoms specialist, the promises of these technologies coming together at scale are still quite some time away from us. Nevertheless, I believe no company with serious digital transformation plan can afford to not have them on the midterm radar of digital transformation technology catalysts.

Quantum Technology   Quantum is expected to transform technology in major areas like computing, sensing, and communications. Quantum computing can solve advanced problems by leveraging quantum phenomena to process information at unprecedented speeds. Quantum communication applies quantum mechanics to create ultra-secure communication networks that should be able to detect any tampering. For sensing, quantum sensing devices are expected to be much more precise than conventional sensors with promising use cases in many sectors.

Exponential Intelligence   The next generation of AI aims to generate a better understanding of human emotion and intent. This has already been proven to be possible, for example, when spotting emotions in client service calls and, in case of anger, routing the respective customer to conflict-trained agents. On top of that, "Soon, these technologies will be able to look at a statistical correlation and, much like the human brain, determine if it makes sense or if it is just a random feature of the supporting data that has no intrinsic meaning" (Buchholz and Briggs 2022).

Ambient Computing Ambient computing will make technology ubiquitous in all our lives. High-performing digital assistants based on an array of sensors, voice recognition, analytics, and exponential intelligence capabilities will be able to accompany you all day: "Augmenting an individual's physical experience with digital information will be another major dimension of life beyond the glass. Researchers and entrepreneurs alike are already exploring possibilities for using smart contact lenses and even implanted brain chips to augment our senses and (literally) read our minds. Think about it: Why wouldn't it be natural to look at the sun and see how many hours until sunset?" (Buchholz and Briggs 2022).

Obviously for these future speculative technologies a discussion of specific accelerators and decelerators does not (yet) make any sense. However, by now you should have enough information from the already-developed technologies to understand what you need to watch out for when they become relevant.

Workforce Skills Digital transformation discussions have recently been accompanied by a focus on supply-side workforce skills in general and the importance of the specific capabilities of digital talent (Kane et al. 2017). The utilization of these skills as a supply-side catalyst is seen as a key success factor for digital transformations to have a positive outcome in terms of high digital maturity (Kane et al. 2017). As such, they can even be an outcome themselves, allowing reductions in energy levels when iteratively addressing further transformation steps in scope. Any "agile" or "hybrid" transformation process (reaction mechanism), as explained later, requires these skill sets to succeed. An actual in-depth analysis of digital talent would be worth a book by itself. For our discussion however, we can summarize it simply this way: In order to succeed in the "Fourth Industrial Revolution" as fleshed out by Schwab, the relevance of supply of higher-order cognitive skills (WorldBank 2018), systems skills, and complex-problem-solving skills will by far outgrow previously more crucial physical skills or technical abilities (Schwab 2017). This is even more true given the impact of "cognitive algorithms, robotic process automation, and predictive analytics tools" to help workers to "spend more of their time on nuanced, complex cases with an opportunity to more directly" (Briggs 2019, p. 7).

Therefore, it should be very clear why almost every technology catalyst discussed earlier also had a strong skill element in its accelerator and decelerator discussion. In fact, getting the capability question right is probably now one of the most crucial elements of any digital transformation: It will become an important driver for any digital transformation business case.

Workforce Skills Payday Accelerators   A sufficiently skilled workforce will bring the ability to apply new ways of working in a much more agile and therefore higher-paced way, which can allow faster time to market of new products and customer-centric journey improvements with potentially positive implications on revenues. At the same time, previously untapped sources of creativity can be leveraged with potential differentiation upsides. Furthermore, in an idealized team of digitally up-skilled teams the need for management overhead can be decreasing. Finally, more flexible working can lead to easier integration with skills outside of the firm—that is, from a wider ecosystem—potentially at lower fixed cost.

Workforce Skills Payday Decelerators   However, the benefits of a more capable workforce also come at a price. Investments into establishing an end-to-end new way of working are required to avoid that the wave of accelerators just discussed is not hitting the breakwater of less-skilled units and thus losing all its power (it might be the IT department in an agile business team or the other way around—the business team in a DevOps skilled development team). Because the skills to set up this new governance and ignite the spark of overall cultural change are usually not sufficiently scaled in-house, external agile change coaches usually need to be hired and kept on board at substantial cost for a significant period until the end-to-end transition has been achieved. Especially in this transition period, this implies that management has less control of what is happening, a dangerous risk, especially if this occurs parallel to a large-scale technical implementation program. It goes without saying that, given the scarcity of the capabilities just explained, additional retention cost and higher salaries must be considered, because the war for such talent is all over the place and loyalty of skilled workers falters, especially when they mostly work in digital-only environments without proper attachment to a brand or a team or even less a manager. This and additional accelerators and decelerators will be explained in more detail later, when addressing the catalyst of new workforce expectations.

Table 4.9 clusters some key accelerators and decelerators for workforce skills depending on their typical tangibility.

Abundance of Funding and New Financing Vehicles

Not surprisingly, many of the described developments under the digital transformation umbrella would not have been/will not be possible if not for "vast amounts of funding" (Andal-Ancion, Cartwright, and Yip 2003, p. 1)

TABLE 4.9 Workforce skills payday drivers.

Drivers	Less Tangible	More Tangible
Accelerators	• Increased speed/time to market • More creativity • Less overhead	• Easier integration of ecosystem, less fixed workforce cost
Decelerators	• External cost (coaches, consultants) to establish new end-to-end way of working • Culture change management cost • Loss of control/cost to establish new types of governance	• Higher salaries • Increased retention cost

and new innovative financing and governance vehicles (Rodriguez-Ramos 2018) to fuel the development. Sufficient supply-side abundance of funding and new financing vehicles (e.g., via blockchain or crowd-investing) therefore is another key catalyst of any sizable digital transformation. While this was obvious in the early days for the now-established platform companies like Amazon and Google, the new vehicles still mostly have reduced the funding barrier for start-ups and early-stage ventures (Gale and Aarons 2017).

Nevertheless, adequate funding is as important or even more important for established firm digital transformations in scope of this book. The difference is that the hype of disruptive funding mechanisms like initial coin offerings (ICOs) as a new form of "digital governance" (Rodriguez-Ramos 2018, pp. 121–152) still plays much less of a role in large firm digital investments and governance set-ups. Widely discussed examples like Tesla's investments in bitcoins are still far outside the norm. In more established firms, the supply of funding for digital innovation and transformation is usually either channeled via corporate venture mechanisms (Benson and Ziedonis 2009; Keil, McGrath, and Tukiainen 2009), built on "the catalyzing effect of corporate entrepreneurship" (Yunis, Tarhini, and Kassar 2018, p. 344), or by establishing joint ventures with other firms (for example, to invest in aggressive fiber infrastructure rollouts) in the ecosystem or financial investors. All are substantial fields of research by themselves. For larger-scale digital transformations, this comes hand in hand with the mundane challenges of controlling the optimal use of invested capital on an operational level and in normal corporate planning cycles (Baumöl 2016; Schönbohm and Egle 2017).

While the accelerators and decelerators of this catalyst are often hard to quantify, they still represent an important element of any digital transformation.

Funding Payday Accelerators First and foremost, new sources of funding open additional sources of capital that can potentially be even cheaper than traditional funding mechanisms (e.g., hybrid sustainability bonds or joint venture investments to generate additional sources of capital at improved financing cost) if the respective financing or partnering entities for one reason or the other want to push these instruments into the market or establish a foothold in a certain segment. These financing savings can be quantified and considered as an accelerator. At the same time, the established new financing structures allow for different risk appetites and acceptance in the new portfolio, which can also be beneficial in more disruptive digital transformation journeys.

Funding Payday Decelerators Unfortunately, the new funding mechanisms do not come without any risks. Beyond the obvious fluctuations in any bitcoin-related instrument any partnership and new source of capital might lead to very different payback expectations between partners. This can, in my experience, significantly increase the often-hidden cost of governance, as substantial efforts must be made to keep the shareholders but also to some extent other stakeholders aligned on an often less overlapping strategic and financial objective than originally foreseen. This is even more true when the new sources of funding are concentrated on investments outside of the firm's core business, because there these experiments are easier to implement in new firm constructs and unusual capital structures and earn-out models. All these further aspects amplify barriers to reach the goal of a positive impact on the core business later.

Table 4.10 clusters some key accelerators and decelerators for workforce skills depending on their typical tangibility.

TABLE 4.10 Funding payday drivers.

Drivers	Less Tangible	More Tangible
Accelerators	• New sources of capital • Allowed higher risk appetite in the portfolio	• Lower financing cost
Decelerators	• Higher governance cost • Increased barriers to transfer to core business	• Increased risks

CHAPTER 5

Demand-Side and Overarching Catalysts

Customers and Workforces Have Changed; From You They Expect Nothing Less

O n the **demand**-side, three catalysts from very different directions can influence your digital transformation.
We see:

- Customer needs
- Workforce expectations
- Spending patterns

And remember, there is also one **overarching** element: blurring industry boundaries.

Demand-Side Customer Needs

If one needs to single out the publicly most visible catalyst category driving the need for firms to digitally transform, it would be the apparent changes in business-to-consumer (B2C) and in business-to-business (B2B) demand (Schwab 2017). However, these catalytic changes in demand-side customer needs, related workforce expectations and spending patterns are never iso-lated. They are closely intertwined with or even triggered by what the supply-side technology capabilities, workforce capabilities, and abundant funding mechanisms have to offer.

Demand-side customer needs are shifting. On the B2C side, new demo-graphics are changing customer expectations (Gale and Aarons 2017; Osmundsen, Iden, and Bygstad 2018), and once-successful approaches of segmentation are no longer relevant (Schwab 2017). Even though it is never advisable to work with over-generalized segments in a more and more indi-vidualized world, new customers (first the Millennials or Generation Y, and now the so-called Generation Z), best characterized as digital natives who grew up in a world of increasing digital access (Gale and Aarons 2017), now often set new consumer trends (Raskino and Waller 2015). Early in life they learned different social behaviors, and became accustomed to business trans-actions customized to their needs (Westerman, Bonnet, and McAfee 2014). Online access to "more and more information is leveling the playing field in any market" (Gale and Aarons 2017, pp. 37–39); this information abundance (Bharadwaj et al. 2013) can tremendously reduce search cost for your custom-ers and limit their contracting risk (Andal-Ancion, Cartwright, and Yip 2003).

Therefore in general, customers' expectations have increased substan-tially (Fitzgerald et al. 2013; Westerman, Bonnet, and McAfee 2014) over the past 10 years. Strongly empowered by digital technology (Gimpel 2015) cus-tomers' expectations have shifted (Fitzgerald et al. 2013; Westerman, Bonnet, and McAfee 2014). They evolve from a mass market segment to a dynamic network of influencers (Rogers 2016). For their trusted companies, they place significantly more emphasis on digital-born firm traits like digital interaction capability or what is called "electronic deliverability" (Andal-Ancion, Cart-wright, and Yip 2003, p. 37) in general. For them, this includes full trans-parency and seamless omnichannel experiences. In addition, they are often assumed to have a different view on sharing asset ownership especially in cities (Schwab 2017), shared transportation (Rodriguez-Ramos 2018) being the most mentioned example. This also influences their spending patterns, which will be elaborated later. In any case, long-term loyalty of customers to firms is at risk: "Millennials and Generation Zs, in general, will patronize and support companies that align with their values; many say they will not hesi-tate to lessen or end relationships when they disagree with companies' busi-ness practices, values, or political leanings" (Deloitte 2019, p. 1). Customers can and will switch and choose services instantly, every time and everywhere (Schwab 2017). This puts significant pressure on firms to adjust not only their customer-facing front-ends but also all other affected parts of their operations.

The same is even more relevant for B2B. Blurring industry and firm boundaries (Raskino and Waller 2015; as described later), maximum trans-parency in a digital supply chain and in addition instantaneous ways of processing transactions (Schwab 2017) impose significant demand-side pressure on firms. They need to adjust their vertically integrated model to a B2B2C model without "endangering third-party distribution models which have been successful for years" (Westerman, Bonnet, and McAfee 2014, p. 88).

Many accelerators, but also—on the downside—decelerators evolve from this key catalyst.

Customer Needs Payday Accelerators

The changing customer behavior and more digital transaction can allow for significantly lower cost to serve if addressed with the proper technical solutions and customer-centric journeys, either because less human interaction is needed or because any upcoming issues or requirements are resolved at much higher speeds based on what has been enabled by the already discussed technology catalysts. At the same time, the new expectations of customers can also open a window of opportunity of getting to know them even better for potential cross- and upselling because their expectations also lead to the necessity to share more of their habits in concrete data. This can be analyzed and leveraged with the right technology and processes. New sources of revenue are becoming possible.

Customer Needs Payday Decelerators

The other side of the coin is that customers are either less loyal or additional cost must be planned in to ensure that loyalty is remaining at similar levels. With brand "stickiness" evaporating, other, usually direct cash out implying retention cost, must be planned in, or substantial investments are required to generate the purpose customers are looking for to remain attached to a brand as such.

Table 5.1 clusters some key accelerators and decelerators for customer needs depending on their typical tangibility.

TABLE 5.1 Customer needs payday drivers.

Drivers	Less Tangible	More Tangible
Accelerators	Increased willingness to share information as a data source for cross- and upselling	Lower cost to serve
Decelerators	Need for purpose-driven investments	Increased retention cost

Demand-Side Workforce Expectations

Quite logically, the shift in consumer demographics just described goes hand in hand with rapidly changing demand-side workforce expectations (Deloitte 2018; Kane 2019). Firms need to spend significant effort to understand and act upon what the required digitally skilled workforce of their future (Millennials and even more different Generation Z) wants. At least the sought-after subsegment with social and creative skills plus the ability to make decisions under uncertainty (Schwab 2017) have different expectations from their corporate employers to convince them, for example, in competition with the "gig economy" (Deloitte 2019, p. 15). For these subsegments, while financial benefits are still most important, factors like a "positive workplace culture," "flexibility," "opportunities for continuous learning," and "well-being" are of high relevance (Deloitte 2018, p. 18). This is a significant challenge for companies aiming to leverage this digital talent market in the future (Gale and Aarons 2017; Kane 2019) and adds a range of decelerating factors to your digital transformation. These can be balanced with a select few accelerators.

Workforce Expectations Payday Accelerators

The new type of workforce, however, does not only come at a different cost structure (less need of office space, less travel), but it also provides some key benefits as accelerators for your digital transformation payday. First, you suddenly have additional levers for attracting top talent in your toolbox at potentially lower recruiting cost. You can tap new sources of digital talent previously only available from consultants or freelance networks, who have been difficult to integrate into your permanent staff. This provides a much more flexible workforce cost structure in all areas in which you decide to partner with/employ these talents. They can stay with you for a key project, for a certain timeframe or even, when you manage to gain their affection and loyalty, be integrated into your permanent teams.

Workforce Expectations Payday Decelerators

On the other side, you cannot just integrate this new type of employees without any cost. You will need to invest in new career paths, lifecycle working models, more creative studio-like office locations, purpose and cultural change management campaigns, heavy reskilling, and continuous learning programs on state-of-the-art digital topics and ongoing on- and offboarding cost when they join your projects or move on to different tasks. Still, these skills will continue to be extremely scarce, so do not be surprised when an increasing salary cost bubble

TABLE 5.2   Workforce expectations payday drivers.

Drivers	Less Tangible	More Tangible
Accelerators	• More flexible workforce cost structure	• Reduced consulting cost for digital talent • Lower office cost • Lower travel cost
Decelerators	• Investments in new career paths, lifecycle working models, more creative studio-like office locations, purpose and change management campaigns, heavy reskilling, and continuous learning programs on state-of-the-art digital topics • Ongoing on- and offboarding cost when they join your projects or move on to different tasks	• Salary increases

is building. Already today, special expertise skilled staff can ask for salaries at levels close to a bursting bubble from a business perspective and they get them, because there always seems to be someone in the market who is willing to pay.

Table 5.2 clusters some key accelerators and decelerators for workforce expectations depending on their typical tangibility.

Demand-Side Spending Patterns

One catalyst, closely intertwined with the aforementioned changing customer needs and the later-described blurring industry boundaries, is the fact that—at least on the consumer side—demand-side spending patterns seem to converge. This cannot be analyzed in detail here but needs to be mentioned since it would be an important separate topic for more in-depth research. "The affordances of pervasive digital technology enable innovations of convergence in a number of ways." They bring previously separate user experiences together and "create convergence because digital technology is increasingly embedded into previously non-digital physical artefacts, creating so-called 'smart' products and tools" (Youngjin et al. 2012, p. 1399). The underlying assumption is that previously distinct budgets in consumption are shifting in the wake of user-experience-driven convergence. Examples include telecoms,

over-the-top (OTT) media companies converging in quadruple-play propositions or that budgets for personal transportation are reallocated with the rise
of the sharing economy because the preference for access over ownership and
tendency for digital-physical subscription models increases toward a tipping
point (Schwab 2017). The shift allows firms to capture demand beyond their
classical vertical industry silos.

This leads to an accelerator and a decelerator becoming relevant, seemingly high level and almost impossible to properly estimate, but still of crucial
relevance for many digital transformation efforts.

Spending Patterns Payday Accelerators

On the accelerator side there is only one disruptive change to be harvested
when relevant in your markets. With converging spending patterns you can,
if done properly, suddenly search for ways to increase your share of wallet, or
even better put, address wallets beforehand outside of your reach.

Spending Patterns Payday Decelerators

The downside is that as spending patterns converge, it becomes less clear
which wallets are addressable, meaning you will face more competition. This
does not necessarily only mean price competition but, probably even more
important, the competition for the attention of your much less loyal and no
matter which generalized segment they might be believed to belong to, fully
individualized customers. It should be obvious that, without adjusted marketing and sales spending, and the technology capabilities that enable you doing
so and targeting them properly, this can become a substantial downside if
your competitors are more advanced than you in this space.

Table 5.3 clusters some key accelerators and decelerators for workforce
capabilities depending on their typical tangibility.

TABLE 5.3 Spending patterns payday drivers.

Drivers	Less Tangible	More Tangible
Accelerators	Increase share of wallet	None
Decelerators	Cost due to increased competition for attention (sales, marketing and technology costs)	None

Overarching: Industry Barriers Are Blurring, Whether You Like It or Not

As demonstrated by firms like Uber for the taxi and Airbnb for the hotel industry, one aggregated characteristic of digital transformation, driven by the aforementioned supply- as well as demand-side factors, is the blurring of industry boundaries (Rogers 2016) enabled by digital substitution (Raskino and Waller 2015): Convergence brings together previously separate industries (Youngjin et al. 2012). "New competitors [. . .] built to be digital from day one" (Gale and Aarons 2017, pp. 40–42) leverage new business approaches eroding or even disrupting barriers that once protected one industry from others. They trigger a disaggregation of traditional industry silos (Schwab 2017), creating new competitive threats but also enabling previously unthinkable partnerships along the value chain (Rogers 2016). Especially for incumbent firms, this combined supply- and demand-side driver is often a key catalyst for venturing into a digital transformation. They are urged to consider a platform play (Bughin and Catlin 2017) to establish new multisided platforms before someone else manages to enter and control their current core business (Westerman, Bonnet, and McAfee 2014). Based on the notion that literally all companies become technology companies, former barriers of entry are less applicable (Raskino and Waller 2015). These catalytic effects of eroding barriers include (heavy) asset infrastructure and products and services that can be digitally substituted and access to capital that can be sourced via new digital-born financing vehicles. It continues with regulation, which can be circumvented by digitalization; and technology that can be copied and branded, which is easily displaced. It concludes with industry experience that can be replaced by (big) data insights and customer loyalty that can and will be switched in real-time (Raskino and Waller 2015; Westerman, Bonnet, and McAfee 2014).

What seems to be an abstract catalyst can have quite drastic accelerating and decelerating effects on your digital transformation payday, which are admittedly almost impossible to quantify properly.

Industry Barrier Payday Accelerators

Eroding industry barriers simply present the opportunity to diversify and attack the business models of competitors that previously had easier means to protect their asset base.

Industry Barrier Payday Decelerators

Unfortunately, attacking other industries' profit pools comes at a price, no matter how low you believe the previously relevant entrance barriers have become. It requires you to invest in specific skills, know-how for verticalization, and make-or-buy decisions. You must pick your battles. One example is the whole idea of the Internet of Things (IoT) and 5G-based business models. These are easily defined and put in presentations for sexy individual use cases but scaling them to really matter in a certain industry from the outside is a totally different thing. Without a carefully planned and prepared proper ecosystem approach, you will end up investing substantially to do it all by yourself, be it in product design, solutions sales, and many more. You cannot attack too many other industries at once and risk spreading yourself too thin without building on a clear strategy and plan for winning in the new marketplace.

Table 5.4 clusters some key accelerators and decelerators for workforce capabilities depending on their typical tangibility.

TABLE 5.4 Overarching payday drivers.

Drivers	Less Tangible	More Tangible
Accelerators	New sources of revenue	None
Decelerators	Verticalization cost (skills, ecosystem building, sales and marketing cost)	None

CHAPTER 6

Reactants/Scope

Make Sure Your Investments End Up Transforming Your Core

As visible from all practitioners' and academics' work so far, there is some agreement that the core of your firm needs to be the ultimate reactant in the scope for any digital transformation to matter at scale. Without transforming the core, you will most certainly end up just digital-washing your company at the edges of your business with an operating model without a strategy and transformation for sustainable success. Still, this does not mean that this scope decision could not be a necessary and winning strategy from a portfolio perspective if there is clarity that your core business is in any way going to be disrupted beyond salvation. Other than prescriptive advice, there is no common structure for further breaking down the scope (what we call the reactants in our framework), in terms of relevance and digital transformation payday acceleration or deceleration.

To make this dimension of digital transformation more tangible, I find it to be very helpful to cluster the potential reactants in a simple two-by-two scope matrix as sketched in Figure 6.1. One axis defines whether you focus on your existing, incremental, or new markets and customers; the other decides whether you leverage existing, incremental, or new products and relevant assets. Depending on how far you go in each of these dimensions, you either transform:

- The **core** of your business, the heart of your current business model and operations,
- The **adjacencies** of your business, which are still closely related to your business model and operations today but leverage incremental add-ons, or
- The **frontier** of your business, which is most often a greenfield approach that has little relation to what you are doing today.

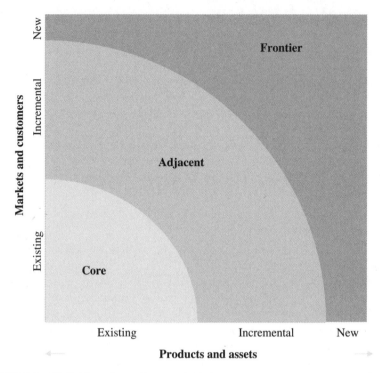

FIGURE 6.1 Digital transformation reactants/scope.

Where and in which order you start heavily depends on what you define in your strategy for winning in the marketplace. This is one of the most important digital transformation choices to make. Obviously, this first requires understanding the digital transformation payday accelerators and decelerators in each of these choices in detail.

Core Business Digital Transformations Are Prone to Failure But Must Be Your Final Goal

Many of the previously described catalysts and reaction mechanisms (described later) can allow firms to radically redesign the way they operate in their core (Westerman and Bonnet 2015). But beware: Putting your core (or, as others call it, the center or legacy business) into focus when first embarking on the digital transformation journey is certainly a daring choice to make. This is even more true for cases in which it is clear that your shareholders mostly "value the company for its current profitability" and not for its future growth potential (Mani, Nandkumar, and Bharadwaj 2018, p. 12). Addressing the core at scale first will potentially make your short-term profitability suffer until envisioned benefits materialize, unfortunately often with a very difficult-to-plan time lag. Any such effort of digital substitution that radically reduces cost in the core business to fund the new game (Charan 2016), standardizing plus focusing on customer experience in the core business first (Weill and Woerner 2018). Leveraging digitally transformed operations therefore requires a "clear, ruthless direction from the center around which projects to scale and in what order" (Baculard 2017, p. 3).

Sounds too abstract? An interesting metaphor to better describe a core-focused digital transformation is an analogy I find very fitting—the analogy to urban planning (Beswick 2017). Core business digital transformation can be compared with the City of Boston's heavy investment into the so-called "Big Dig" project, aiming to replace legacy highway infrastructure with an underground tunnel network. This is very similar to digitally replacing legacy core systems and quite close to the philosophy of enacting real options (Copeland and Antikarov 2001) for later flexibility and growth. What sounds so easy in writing is at the same time the riskiest but also potentially most rewarding lever of any digital transformation. Think of a sizable technology company with literally hundreds of legacy systems with a digital front-end attached with duct tape. Transforming this to a cloud-based best-of-breed technology stack that is headless, microservices-based and evolving based on continuous agile development trains, as explained in Chapter 7, is usually the biggest project these companies have been facing for years. We are talking about a program that moves from an idea to an actualized state that is a multiyear exercise involving hundreds of project participants from inside and outside the organization and hundreds of millions of dollars of financial investments.

Obviously, the digital transformation payday accelerators and decelerators for such a core transformation approach are often hard to derive, due to mostly indirect impacts.

Core Business Transformation Payday Accelerators

In an ideal world in which all digital transformation elements described in this book play seamlessly together, the business case can be easily made to look promising. Without the big capex requirements that needed to be made up-front in the past, you continuously spend to replace legacy systems one by one and quickly negate the obvious negative additional opex impacts by harvesting all the beautiful benefits on the customer, process efficiency, and cost sides (for more on this, see Chapters 4 and 5). Not only that, in your core business's idealized future, all catalysts play their strengths together in a virtuous network to make their sums more than just a simple addition of benefits. And not only that. At the same time, you build a sustainable platform for all your future growth dreams, your real option on your future success. Achieving this is not impossible, but hard work.

Core Business Transformation Payday Decelerators

Unfortunately, my hard-learned experience shows this only works in flashy presentations. Core business transformations often quickly become (usually after six to eight months) like mass heart surgeries in a field hospital, without enough experienced surgeons and with surgical instruments from many different ages. This carries a substantial cost, decelerating your digital transformation payday. Not only do all the decelerators described in Chapters 4 and 5 come into play, but they also amplify each other when working all together, and especially when affecting the heart of your business operations. This starts with the aggregated cost of all the best-of-breed platforms to be implemented—the foreseen and unforeseen integration work between these solutions—and continues with all the external experts needed to design and steer your transformation and ignite the required change in your teams. It ends with the unfortunate fact that is hardly ever factored into business plans—that you have only one team and cannot run business as usual and such a massive program of change at the same time. Core business transformations almost certainly require that you do large-scale backfilling in your staff for as long a time (often years) as the program and your legacy will have to run in parallel. Also, tough choices must be made: Do you wait years for the program to deliver step-by-step what will hopefully be next-generation and market-differentiating features, or do you accept that you need to develop some version of these features already in your legacy systems and thus invest twice, only to scrap one later for good? In any case, always plan with a substantial risk buffer and factor in substantial control and governance cost. And this does not include that programs to achieve this goal often fail on the first and often the second attempt, a fact that is hardly ever factored into business cases to allow management to start fresh each time. While the often-used argument of sunk-cost irrelevancy at that point in time may be economically correct, it can be misleading when you first design your overall digital transformation journey.

Obviously, life is much "easier" when you have no choice, when your core business is in danger of being disrupted or dying and needs to be completely replaced by something utterly new, no matter what. The other question is whether your shareholders are actually willing to give you the funds to start such a binary win-lose journey.

Table 6.1 summarizes the drivers of core business transformation.

TABLE 6.1   **Core business transformation payday drivers.**

Drivers	Less Tangible	More Tangible
Accelerators	• Single benefits of all applied catalysts and reaction mechanisms plus virtuous network of multiplying benefits • Enablers/real options on the core business's future success	• Only escape route in case of core business disruption
Decelerators	• Single decelerators of all applied catalysts and reactions mechanisms plus multiplying complexity cost and interfaces • Risk buffer	• Substantial control and governance cost • Staff backfilling • Parallel double invests (legacy- and new technology/processes/ products) • First-time failure cost

Digital-Adjacent Businesses Will Never Be Family Without Proper Reintegration

Once more following the aforementioned urban-planning analogy, your option of scoping a digital transformation for the adjacent business is comparable to making targeted investments in a few lighthouses outside of the core, serving as a starting point for further transformation. The city of Dubai historically took this approach when first erecting high-rise skyscrapers. The successes of these skyscrapers were the initial spark that spilled over to the rest of the city and took Dubai into a totally different league of modernization within astonishingly few years. In business, the hope is that implementing new online capabilities, introducing data analytics, and launching digital labs or innovation centers to change perception can later enable the core "reinvention that needs to follow" (Beswick 2017, p. 3). The digital extension (Westerman et al. 2011) here comes more from diversification and experimentation with technologies (Bughin and Catlin 2017) rather than addressing challenges the hard way, as explained before in the core business. If you agree with me that in the end you aim to transform your core, you must factor in the substantial cost of planning and successfully executing the later reintegration of your adjacencies into your business. This does not happen automatically.

Other approaches only make sense if your strategy is to become a financial holding company where your adjacencies prosper, and your core business dies off (which better fits to the frontier approach explained in the next section). Over the years, I have seen the effect of a defensive reaction of a core business when the adjacencies are pulled back for scaling. The effect can be so strong, that the adjacent successes are quicker killed than built. Nevertheless, accelerators and decelerators for an adjacent transformation are usually easier to quantify but initially small in scale.

Adjacent Business Transformation Payday Accelerators

The beauty of adjacent businesses is that the accelerators can usually be siloed and clearly defined. Therefore, many of the catalysts and reaction mechanisms as explained in the previous and following chapters are usually first tested in an adjacent model to limit the overall risk. Either by experimenting in existing units (for example, focusing on front-end technologies like the website or the online channel) or in separate companies focusing on new business models (for example, fintech experiments of insurance companies or IoT software spin-offs of telecoms companies). Then you can easily structure and plan the benefits as clear revenue or cost-saving streams along certain accelerators without getting lost in the complexities of multidimensional, functional and timing dependencies. This increases speed and agility, tremendously reduces your risk exposure, and reduces your control and governance cost. It may also produce indirect positive inspiration and spillover benefits for your core business.

Adjacent Business Transformation Payday Decelerators

While the decelerators in a ring-fenced adjacent transformation are usually easier to plan along the catalysts explained earlier, one major decelerator I have seen has been mostly forgotten—namely, the previously described need for a reintegration. Without this, you will never ignite the spark of transformation in the core too, to make it accepted, and you will not learn from the adjacencies' way of doing things and deploy new technology advancements at scale. Typical reasons for failing to do so include the lack of early forced touchpoints (for example, making the core a compulsory customer of the products and services of the adjacent business); a forgotten early technical pre-integration of innovative platforms, underinvesting in cultural alignment

TABLE 6.2   **Adjacent business transformation payday drivers.**

Drivers	Less Tangible	More tangible
Accelerators	• Easier plannability • Risk ring-fencing • Positive indirect benefit spillover	• Speed/agility
Decelerators	• Softer reintegration cost (culture, people, change)	• Technical reintegration cost for scaling in the core

and change programs to avoid the "not invented here syndrome"; clashing cultures; ways of working and different language for similar things; or simply by misunderstanding each other from day one. Without carefully planning this and accepting that you need to do so and invest even more in it than in the first adjacent experiments, I can guarantee you will lose everything.

Table 6.2 summarizes the drivers of adjacent business transformation.

Digital Frontier Businesses Will Not Behave (the Way You Expect)

The most disruptive scope decision for digital transformation is probably completely starting anew and breaking away from the past, which is often most difficult to explain. This is particularly true for established companies in noncrisis situations. Coming back to the urban planning analogy discussed earlier, this is an approach to a new frontier like starting from zero, similar to the city of Shanghai, which built a completely new financial hub. In terms of digital transformation this could imply a greenfield approach, or breakthrough (Westerman et al. 2011), for example, launching new cloud-based systems and migrating only what is of key importance (Beswick 2017). Business benefits for this frontier transformation approach are usually difficult to quantify due to their high volatility of outcomes.

Frontier Business Transformation Payday Accelerators

The benefits of a frontier business transformation are somewhat obvious. As with any new venture, they might take some time to build greenfield, but they are not slowed down by any legacy chains of your core business. Once they are successful, however, they can open completely new revenue streams for you while running on state-of-the art technology and a lean operating model, which does not have to compromise in any dimension on limiting factors from your legacy. Also, you can substantially reduce your control cost vis-à-vis an adjacent transformation, because frontier businesses mostly need to be steered at arm's length to allow them to breathe and succeed without being strangled by large-scale corporate processes.

Frontier Business Transformation Payday Decelerators

What are the decelerating factors of a frontier business transformation? None, at least if you forget the fact that you are not actually transforming anything. Instead, you are creating something completely new, with no limits to what you have done before, and therefore there is also no positive impact whatsoever on your core business. That does not mean that it wouldn't be a good idea if the core business is disrupted or about to be disrupted beyond repair. But if you embark on a frontier transformation to help your core business, you have to expect unexpected behavior and may end up with a core business that must be restructured or divested. Certainly, if you plan to use a frontier business transformation to save your core, you must factor in substantial reintegration investments with even higher barriers in terms of culture, people, technology, and processes, and it becomes even bigger than choosing adjacent reactants as your transformation starting point.

Table 6.3 summarizes the drivers of frontier business transformation.

TABLE 6.3 Frontier business transformation payday drivers.

Drivers	Less Tangible	More Tangible
Accelerators	• Freedom, speed • No legacy chains	• Limited control cost compared to adjacencies (financial holding like)
Decelerators	• Very high reintegration cost if aiming at impacting your core	• Likely no impact on your core if remaining greenfield

CHAPTER 7

Reaction Mechanisms/Process

Agile Done Wrong Can Be a Very Effective Tool of Value Destruction

Instead of working on the required ingredients of a clearly defined winning strategy (Lafley and Martin 2013)—which will be elaborated later—or getting a clearer understanding of the previously discussed reactants/scope, I have recently seen practitioners shifting more and more to focus on reaction mechanisms, that is, the transformation process.

No matter what you may hear or read otherwise, without a clear strategic vision, target state, and scope in mind, this is definitely not a winning idea. All too often, the business buzzwords of the day, like *agile, hybrid, dual-speed*, or *bimodal,* which originated from software and IT development practices, are applied and implemented without proper understanding and without consideration of their potentially dangerous and digital transformation payday decelerating implications, when rolled out incorrectly or incompletely. I can only recommend ensuring you have carefully understood their strengths, limitations, and most importantly their prerequisites for success. Only then can you prevent the root of value destruction from strangling your digital transformation program from the outset.

Waterfall Is Not Dead Yet, But It's About to Be Retired

Before we dig deeper into these process buzzwords, let me make one thing clear based on experience: What is often disrespectfully denounced as waterfall in digital transformation practice today has for a long period right-fully been the most-used approach to manage large and/or complex projects in mature organizations. (Smaller firms can get additional specific guidance at the end of this part, in Chapter 12). Even today waterfall has under certain circumstances lost nothing of its former relevance, even though admittedly such circumstances are often very limited. Life-critical applications that need a formal proof of being error-free (e.g., using mathematical methods) are such an example. There also might be parts/elements in large-scale projects that are implemented using these formal waterfall methods, whereas the rest is done iteratively.

So, do not feel pressured to manage everything based on agile methods. Transformations that have "naturally many interdependencies", "require [a] large amount of coordination" and fit best to mature IT governance frameworks "such as ITIL, CMMI, or COBIT" (Barlow et al. 2011, p. 33). Waterfall elements should not be discarded too early and too arrogantly. On the contrary, certain ingredients of plan-based approaches to steer large-scale digital transformations still have not lost all their appeal when team sizes are large, and the nature of project interdependencies is fully sequential. Still, I have seen many organizations and projects fail despite the fact that all have been certified on Capability Maturity Model Integration (CMMI) Level 5 or higher. Also, some tasks, such as a major revamping of your business application portfolio with significant implied hardware and software changes, might still require a "big bang". For such projects, you cannot just iterate across your typical agile sprints. On the other hand, waterfall approaches obviously quickly show their weaknesses, when interdependencies become more reciprocal (which is the reality of any digital transformation) and smaller team sizes enable higher flexibility (Morton, Stacey, and Mohn 2018).

As the accelerators and decelerators of waterfall transformations should not be news to you based on years of experience, they are only quickly summarized in the following sections and in Table 7.1.

TABLE 7.1 Waterfall transformation payday drivers.

Drivers	Less Tangible	More Tangible
Accelerators	• Prerequisite for large-scale big-bang changes • Limited change, communication, and reskilling cost	• Budget control and plannability • Must have for error-free requirements
Decelerators	• Slower time to market • Plannability of large-scale programs only exists in theory	• Double investments in legacy and to-be systems

Waterfall Transformation Payday Accelerators

The first and foremost digital transformation payday accelerator and the single most important reason that the waterfall concept was originally introduced is still valid: In large-scale sequential big-bang settings with a clear target, you can hope for better ability to plan and manage budget control, lower external and internal communication cost, and more manageable reskilling requirements.

Waterfall Transformation Payday Decelerators

Obviously, all these benefits come at a price. Waterfall requires you to control a large-scale project with hundreds or thousands of team members up-front by doing a proper analysis, planning, and execution. What is missing from this equation are the uncertainties, the changes that occur over time (driven by internal and external factors), and the unmanageable complexity of such projects. No matter what, waterfall approaches will progress slowly until any impacts of what you do become measurable. Even worse, you may end up investing in a target state that is no longer differentiating from competition once you reach it or you will have to invest twice to make that happen, once in your legacy and once in the new technology implemented. Also, in reality, the hope of being able to plan often turns out to be a naive dream. The results are many failed commitments, unrealistic timelines, and false expectations based on planning with very little knowledge.

The Light Side and the Dark Side of Agile Transformation

So is agile the lighting in a bottle you can use to shake-up a long, slow-moving waterfall? The original idea of agile in software development first introduced iterative development, frequent customer feedback, timely releases, and rigorously testing (Cao et al. 2009). In the broader digital transformation context, agile concepts were first leveraged in the context of agile organization discussions (Roberts and Grover 2012), "organizing for agility" (Kane 2019, pp. 169–178), or agile strategy (Morton, Stacey, and Mohn 2018), where iterative testing and prototyping (Rogers 2016) are usually a defining element. Other proponents see agile thinking as success critical to increase the responsiveness of the IT function (Haffke, Kalgovas, and Benlian 2017), improve business/IT alignment toward a business IT (Briggs 2019), and recommend it

TABLE 7.2 Agile transformation payday drivers.

Drivers	Less Tangible	More Tangible
Accelerators	• Better value-based prioritization • Quicker reactions to market-driven scope changes • Faster time to market • More manageable risk because of more intermediate checkpoints • Simpler decision processes but most importantly	• Reduced development times at higher quality and therefore less cost
Decelerators	• End-to-end agile concept and target operating model invest • Invest in new funding mechanisms and governance processes • Agile vendor management approach	• Training and reskilling cost • Recruiting cost • External expert cost (consulting, agile coaches) • Communication and change management cost

at least for smaller teams in reciprocal interdependency environments (Barlow et al. 2011).

While nowadays, agile methods and thinking (e.g., release planning sprint, planning, and daily scrums) have become a standard element and vocabulary of daily transformation practice, many organizations are still struggling to reap the benefits when not all accelerators and decelerators are carefully taken into consideration early on. These are summarized in Table 7.2 and the following sections.

Agile Transformation Payday Accelerators

The advertised, digital transformation payday accelerating benefits of agile are clear. I am sure they have been presented to you in flashy presentations many times. If done the right way, you can expect reduced cost, reduced development times, higher quality, and faster time to market, and you will enjoy a more manageable risk portfolio because of more intermediate checkpoints and simpler decision processes, but most importantly, you leverage the opportunity for a much better value-based prioritization with quicker reactions to market-driven scope changes. This is especially true when agile acts as an amplifier together with the previously described reactant/scope choices. Obviously agile is easier to implement at the edges of your business, that is, in frontier or adjacent business models (e.g., a new product) or functionally (e.g., in front-end website design).

Agile Transformation Payday Decelerators

While all this might sound like a no-regret decision to implement all over the place in your digital transformation, problems typically materialize when agile is taken from such a generic concept to a real-life program at scale and beyond the edges of your organization. Sadly, if not set up correctly from the beginning, most agile transformations will be doomed to fail before they even start.

First and foremost, you cannot start your agile transformation on the go. People often think they can transform the organization and their staff just by doing the program in an agile way. The concepts initially seem to be so simple that this must be possible. But it is not—it requires a paradigm change that has to be managed on its own, and, more importantly, it has to be seen as a real change effort. You must invest heavily in your customized end-to-end agile target operating model up-front or you can be sure of value-destroying breakwater barriers popping up whenever agile methods and traditional management and governance principles still in place coincide. This can be a beautifully designed DevOps concept in IT being hindered by traditional business-requirement thinking on the business side or the other way round—an agile product design process in business hitting the wall of traditional legacy release trains in IT. And all that at substantial governance cost to steer around such inefficiencies. At the same time, if your agile team is not set up properly, you can expect to face dramatic productivity declines.

If you do not set the scene in advance of establishing new funding models beyond classical business-case thinking (as already explained in the catalyst section of this book, Chapters 4 and 5), classical return-on-investment (ROI) discussion will make agile development speeds much less fertile. Stakeholders on all levels must be educated to let go of some up-front certainty and agree on investments as an iterative test-and-learn approach, in which proof-of-concept, prototype, minimum viable product (MVP), and regular iterations are standard.

Agile also requires a far more professional, timely, and customized risk management compared to a waterfall environment. In a waterfall transformation, monthly risk meetings and the typical risk-impact matrices, combined with a mitigation list and log are typically enough. In agile, the risks need to be addressed much faster—in days or hours—if you do not want their implications to be felt in real-life production immediately thereafter.

It is critical to ensure that everyone in your program and in your firm is adequately skilled in the language, methods, and implications of agile. This will definitely require sizable investments in building up and freeing resources with sufficient dedication to be staffed on your transformation. Factoring in a potential major disruption in your business-as-usual becomes inevitable. This needs to be mitigated early with investments in backfilling of resources. You must build new staffing mechanisms that assign and properly onboard cross-functional experts to agile sprints. It also needs to ensure continuity from one sprint to the next and from one project to the other for stability and regular velocity. This often also requires inflexible role boundaries to be removed. If you still structure roles into very strict categories, then you will never have an agile team who can problem solve by itself.

To make these siloed agile teams perform, you will need to continuously invest in agile coaches and training, change management, and—assuming a limited agile expertise in your workforce—recruitment of agile specialists on all levels. No matter what you believe today, you will likely end up paying sizable sums for onboarding external agile skills over the course of your transformation. The same is true for new tooling and automated solutions now available—for example, automated build and deployment tools or better work organization software.

All in all, there are fun elements of agile such as constant prioritization of requirements and quick adaptations. These usually get implemented very fast. But there are also hard things like the commitment of teams (no over-ruling . . . no pressure!), a more democratic organization (yes, you need to talk to each other), and open communication and transparency (something to learn from). The last one, in particular, will help you avoid the dark side of agile methods, which can encourage errors or weaknesses to be hidden for as long as possible, potentially compromising the entire undertaking. The more difficult aspects of agile are not implemented and this omission occurs with a relaxing of traditional methods of planning and execution. The result is chaos. The more difficult elements of agile often are the corrective elements with respect to planning and progress.

Obviously, as a downside of the discussed flexibility, outcomes are also much less predictable and never reach any long-term stability, even though you could argue that the predictability of waterfall processes often turns out to be a myth. Even if you see this statement of clear predictability in a presentation, this luring sense of security of outcomes is not necessarily tangible. Predictability is often merely perceived, whereas agile clearly states and communicates that things can change. This is hard to accept in traditional management, in which we need clear goals, a working plan, and people who make other people stick to this plan—but it is not real. In every "waterfall" project, the plan gets adapted once people can no longer avoid it. But then, months later than it was clear to everyone—you plan countermeasures and backtrack on measures, things that could have been done months before. This means higher communication and change management cost to explain and achieve new ways of working at scale.

All this does not even consider one specific reality of many agile transformations: They do not happen in isolation. They involve a wide range of vendors and partners working together in the same team. This will definitely create an agile tower of babel between suppliers, in which substantial investments must be planned to align everyone on the same language and install new agile vendor-management principles, as classical steering methods will no longer work. You might be chained into fixed-price or fixed-outcome contracts, or your partners might not yet be adequately trained in agile delivery models. Working with vendors who are unwilling to or incapable of joining your agile transformation can be a serious risk. You will need to ask for new contracts or even switch your suppliers.

Hybrid Transformation Is Complex, But You Must Face It as Reality

As usual, the first approach taken to address the weaknesses of two opposite concepts in practice is trying to define a compromise. For digital transformation practice, this is usually summarized under the name of "hybrid" or "bimodal" (Haffke, Kalgovas, and Benlian 2017). I would usually recommend it in all cases, when large team sizes and complexity meet a rather reciprocal and volatile catalyst-driven environment and plan-based methods that only have limited stability. Hybrid practices can amend the shortcomings of pure agile methods reintroducing some key elements of the "waterfall" approach. This includes principles like up-front design, channeled customer feedback, flat hierarchies with controlled empowerment, and so forth (Barlow et al. 2011). Plan-based approaches remain part of the solution, because just expanding IT's and business's mandate to more flexibility "does not reduce the need for it to continue its operational responsibilities for delivering reliable, scalable, secure and efficient enterprise systems" (Haffke, Kalgovas, and Benlian 2017, p. 103). For such a hybrid digital transformation process, managers need to have both exploration and exploitation of their firms' resources in mind (Hess et al. 2016).

However, compromises hybrid methods do not only provide means to accelerate your digital transformation payday, they also come at a cost in terms of decelerators. These are summarized in Table 7.3 and are explored more fully in the following sections.

TABLE 7.3 Hybrid transformation payday drivers.

Drivers	Less Tangible	More Tangible
Accelerators	Best from agile and waterfall	Increased control
Decelerators	All decelerators of agile plus additional complexity cost	

Hybrid Transformation Payday Accelerators

Compared to fully "agile" approaches, "hybrid" transformation would admittedly slow you down, but it gives you some more control than pure agile. It also works at much better speeds than waterfall and includes all the benefits of easier communication and somewhat better ability to plan. This is because when you can provide at least a high-level roadmap with some lighthouse milestones over a multiyear digital transformation program, you give everyone—not only your teams but also your external stakeholders—a fixture they can relate to in an often seemingly chaotic agile transformation development with many small successes and failures.

Hybrid Transformation Payday Decelerators

Any hybrid approach makes your agile transformation elements ultimately less clean, and thus adds complexity cost in terms of interfaces and so on, which is the number-one reason for slower speed compared to pure textbook agile transformations. Agile is hard to implement effectively, and often a hybrid approach is more acceptable for managers to try it. And actually, the issues can be overcome. The real hybrid nature of agile is implicit in the process. There is no need for everything to be agile around an agile project. There are proven ways to interconnect an agile method with a traditional waterfall environment, and because agile is flexible, it can quickly adapt to any shortcomings of waterfall projects used to deliver the agile project. It's as simple as shifting the milestone. Unfortunately, hybrid does not mean that you save on any of the decelerators (training, change, and communication cost, just to name a few) explained in the agile and waterfall sections. Quite the opposite, the additional layer of complexity will amplify many of these negative effects and add additional breakwaters at interfaces whenever the two methods are playing together.

CHAPTER 8

Outcomes

Digital Transformation KPIs Are Worth the Pain and Resistance

F inally, we have arrived at the key framework element that is closest to the digital transformation value impact we are looking for—namely, outcomes. As the metaphorical products of our digital transformation reaction process, these outcomes are structured in two different categories with very distinct characteristics: (1) subjective digital maturity status, and (2) objective digital impact information.

The first category, *subjective digital maturity status*, includes all digital transformation products, which are usually not observable by external shareholders and stakeholders without any insider knowledge. This category forms the basis of most digital transformation research, and it mostly builds on the already extensively explained—and criticized—maturity models from Part I of this book. Unfortunately, this category is of little further value for any useful empirical findings because, practically, under the assumption of efficient capital markets (Fama 1970), external observers of digital transformation processes do not have access to or cannot rely on this information. Still, many firms and advisors will happily choose to take the simple path of subjective measurement in many different flavors, mostly to get at least some grasp of where they stand in relation to a predefined internal maturity goal and build a "burning platform" to provoke some reaction on an abstract, functional, or aggregate level in internal digital transformation strategy presentations.

As explained in Chapter 2, you should under no circumstances accept this as an easy way out. With some hard work and very likely against many organizational resistances, you can reach a higher level of understanding based on working with more concrete outcomes. For this you need to leverage the second category, *objective digital impact information*. This must take the form of specific, measurable, accepted, realistic, terminated (SMART) digital transformation descriptions (Kawohl and Hüpel 2018). Fortunately, when assumed to be value relevant, capital market–listed firms must publish or announce them in the wake of their legally obligated or voluntary external communication.

But research insights on these observable and therefore more objective digital transformation outcomes are scarce, even though they would be highly relevant and useful for us.

To give you an overview of some current state-of-the-art research, the following summary table (Table 8.1) tries to cluster objective digital transformation impact products (that is outcome proxies and outcomes) in three groups with increasing tangibility:

1. Replicable references
2. Operational KPIs
3. Financial KPIs

Obviously, every product/outcome can be the materialization of the digital transformation accelerators and decelerators explained in detail in the previous chapters, depending on its expressed impact or trend versus the previous status quo. Table 8.1 provides examples of each category and demonstrates the apparent focus in existing digital transformation research on replicable references and to some extent operational key performance indicators (KPIs) as outcome proxies. Research on real outcomes, that is financial KPIs, is almost nonexistent. No matter which of these categories we look into, we unfortunately find one thing that is mostly missing—the link to externally measurable market value, a key objective of this book.

Finally, a word of advice before you continue this chapter: Even when you believe it cannot be replicated or even measured in official reports, the impact of digital transformation is often still observable, by what people do and what they say they do about it beyond official reporting.

Replicable References as a Proxy to Measure the So Far Unmeasured

Replicable references, our first objective outcome category in digital, can take many different forms and it seems to continue to grow in number and frequency (Kawohl and Hüpel 2018). Stories on firm-wide digital transformation lighthouse efforts have therefore at least raised some initial attention in digital transformation research.

Replicable References: Advancements

Replicable references have one main advantage. They have turned out to be the easiest to capture as an outcome proxy. They thus help to significantly

TABLE 8.1 Product/outcome clusters depending on their tangibility.

Outcome/Product Proxies/Outcomes	Replicable References	Operational KPIs	Financial KPIs
Tangibility	Least tangible	More tangible	Most tangible
Examples	Mentions/statements on digital transformation lighthouse activities, projects, pilots, etc.	Key performance indicators (KPIs), for example, on customer experience or employee experience (for example, net promoter scores/NPS, CSI, digital first ratios), productivity parameters (for example, self-service ratios, first resolution rates, automated transaction shares)	Directly linked digital transformation data from P&L statements, balance sheets, cash flow statements (for example, digital business revenues or cloud cost)
Impact direction (acceleration or deceleration)	Can be accelerator (positive impact) or decelerator (negative impact)	Can be accelerators (positive impact) or decelerators (negative impact) depending on their incremental effect	Can be accelerators (positive impact) or decelerators (negative impact) depending on their type (for example, revenues versus cost)
Relevant digital transformation research	Some: • Text analysis of digital orientation (Beutel 2018): Empirically positive relationship to value demonstrated	Little: • Intangibles like customer satisfaction, quality, processes, customer relationships, quality of human capital, etc. driving value in digital (White 2016): Relationship to value not analyzed	Very little: Indirectly, as one criterion in qualitative text analysis (Kawohl and Hüpel 2018), could fit into "replicable references" category: Relationship to value not analyzed

(continued)

TABLE 8.1 (continued)

Outcome/Product Proxies/Outcomes	Replicable References	Operational KPIs	Financial KPIs
	• Text analysis of digital transformation linked to analyst recommendations (Hossnofsky and Junge 2019): Empirically positive relationship to value demonstrated for midstage, but not for early and late stages of analysis • Market value and digital maturity (Zomer, Neely, and Martinez 2018): Reciprocal analysis, companies that increase market value are more digitally mature • Applying real options to digital transformation (Schneider 2018): Conceptual relationship to value demonstrated • Announcements on E-Commerce/digital (Dehning et al. 2004; Subramani and Walden 2001): Empirically positive relationship to value demonstrated • Linking digital activity announcements in reports and investor calls to market value (Chen and Srinivasan 2019): Positive relationship demonstrated	• Steering models based on "management dimensions" like "community, partner, portfolio and resources" for the digital age (Schönbohm and Egle 2017): Relationship to value not analyzed • Dashboards for digital innovation (Mullins and Komisar 2011): Relationship to value not analyzed • Digital innovation/patents (Mani, Nandkumar, and Bharadwaj 2018): Influence of market expectations on the relationship between digital innovation and firm performance	

advance the understanding of firms' digital transformation impact by defining adequate keywords and then applying qualitative text analysis to public announcements (Beutel 2018; Kawohl and Hüpel 2018; Subramani and Walden 2001). They also can form the basis for much deeper insights based on recent advancements in programmatic natural language processing (NLP), such as dependency analysis, to measurable facts or sentiment analysis (see Appendix B).

Replicable References: Limitations

However, given the lack of measurability of these stories and the implied subjectivity of their impact implications, all findings must be taken with a grain of salt. In digital transformation research, few authors took the necessary steps from mere announcement analysis toward working on finding actual value implications (Beutel 2018; Chen and Srinivasan 2019; Hossnofsky and Junge 2019; Subramani and Walden 2001; Zomer, Neely, and Martinez 2018). They all agree that there might be a selection bias when identifying the replicable references, regardless of whether this is done by humans or machines, potentially leaving relevant terms out of sight or adding too many. From my perspective, this concern can be neglected. As a mitigation, I have chosen a mixed approach, which expands a manually defined dictionary, with definite selection bias risk, by a wide range of natural language-processing-based word vectors and a neutral second digital transformation expert's validity check to eliminate bias as much as possible. There could also be a selection bias in the NLP code applied as such and the taken preprocessing steps to clean up the reports (see Appendix B). While measures have been implemented to reduce this bias (e.g., random checks), this concern cannot be fully discarded. Even the sophistication of the implemented NLP code cannot (yet) replace human understanding of implied meaning of the text, eliminate all false positives (e.g., niche company names), and identify dependencies across paragraphs. As the amount of work required to do the same analysis manually is prohibitive, we must live with this concern until future NLP innovation eases the issue. We must also assume that the potential bias to generate the chosen digital transformation proxies in the same way for every observation still makes the results valid enough for careful interpretation. Also, there might be a bias in the data in the sense that the chosen proxy of replicable references does not adequately represent the companies' true digital transformation status (for example, if the chosen companies predominantly report only positive digital transformation experiences, while companies with failures limit their communication about digital transformation). Like Chen and Srinivasan (2019), I do not see this as a major issue, given the assumption that the actual success of digital transformation efforts will not be visible at the time of disclosure anyways, and therefore likely does not play a major role in influencing disclosure

TABLE 8.2 Replicable references advancements and limitations.

Advancements	Limitations
• Can be measured with simple textual analysis of comparable annual reports	• Potential selection bias in defining what counts as a replicable reference
• Allows further insights based on recent advancements of programmatic natural language processing (like dependencies to measurable facts, sentiments, and much more)	• Potential selection bias in any applied programmatic natural language processing (NLP)
• Accepted mitigations against replicable references selection bias have already been developed	• Potential bias in what companies decide to publish on their digital transformation impacts

decisions. Nevertheless, this concern could hint at a future research direction in terms of multidimensional proxies, including not only the proxy, but also the one-to-one sentiment associated with this single proxy.

Table 8.2 summarizes the advancements and limitations of replicable references.

Operational KPIs Are Elusive, But Better Than Nothing

Our second objective digital transformation impact category, *operational KPIs,* like data on customer experience perception or employee experience (for example, net promoter scores or similar KPIs), or operational productivity parameters (for example, self-service rates in customer service) is very common in daily management practice. Nevertheless it has found few inroads into digital transformation research other than survey-based discussions on relevant intangibles (White 2016) and initial efforts to develop new steering models based on "management dimensions" like "community," "partner," "portfolio," and "resources" (Schönbohm and Egle 2017).

Operational KPIs: Advancements

Operational KPIs can bring one additional advancement to digital transformation value analysis on top of what even the most sophisticated replicable reference analysis can deliver. At least for internal analysis, they can form the baseline for applying more advanced models to estimate the impact on actual

TABLE 8.3   **Operational KPI advancements and limitations.**

Advancements	Limitations
• Ongoing innovations in causal models can provide first insights on the relationship between operational KPIs and financial impacts	• Data not publicly available in any systematic way: Market value impact analysis not possible • Internal analytical cause-and-effect models (on financial KPIs) often still experimental and with substantial noise

financial KPIs. Examples include the recent experiments in developing causal models to estimate the impact of digital customer satisfaction improvement measures on customer experience KPIs like net promoter score (NPS). Based on the results of these analyses, modern approaches then try to find (causal) links of NPS to customer lifetime value.

Operational KPIs: Limitations

The most relevant downside of trying to leverage operational KPIs for systematic digital transformation value analysis is that, externally, this information is not available in any systematic way. This is no surprise, because no definite standards for publishing such information exist (and rightly so). Therefore, the key step of linking operational KPIs to market value is not possible from the outside. This does not mean that internal models should not be in your focus. While still experimental, I highly recommend that you constantly challenge your internal analytics teams and external analytics experts to experiment with this idea of using causal models to stay on top of the current innovation happening in this space. The initial successes I have recently seen from such models are quite promising, even though there is still substantial noise distorting the transparency of effects.

Table 8.3 summarizes the advancements and limitations of operation KPIs.

The Journey to Financial KPIs Is Cumbersome But Worthwhile

Our third category, *financial KPIs* (e.g., data from profit and loss (P&L) statements, balance sheets, cash flow statements) quickly turn out to be the least covered outcome cluster for digital transformation and, even if so, only as part of niche qualitative text analysis (Kawohl and Hüpel 2018). Instead, if at all, financial KPIs are used as the dependent firm performance variable in empirical research. Apparently, conventions on how to capture digital transformation successes have not found their way into research and practice. This is not surprising given the manifold impacts any digital transformation can have on these KPIs.

Financial KPIs: Advancements

The journey to developing a better understanding of the internal value relationships (in the form of value driver trees and causal models) in your organization is still worthwhile. While this might not help you to find a link to market value, as such, you certainly need to go down this route, no matter what internal resistance you might be facing. At a minimum, this will ensure that your internal business cases are not just crystal-ball exercises but based on a learned and constantly improving understanding of what happens with your financials when you work on certain triggers. Furthermore, a systematic recording of financial effects of what you do in digital allows you to decide what you want to communicate in terms of replicable references, and which facts (e.g., revenues from a certain digital product or business model) you want to report to your external shareholders and stakeholders.

Financial KPIs: Limitations

Financial KPIs for digital transformation value impact analysis share the same shortcomings as their foundational IT/IS and innovation predecessors, which, like other non-accounting information, can only be translated into financials using indirect conceptual value-driver trees on revenue and cost line items basis, like extensively demonstrated by Vartanian (2003) for innovation value research. For digital transformation, this idea is still in its infancy so that most likely all experimental analysis will come with substantial bias and even more important noise from other factors (for example, nondigital-related measures or market/competition changes) disturbing a clear transparency on what is happening.

Table 8.4 summarizes the advancements and limitations of financial KPIs.

TABLE 8.4   Financial KPIs advancements and limitations.

Advancements	Limitations
• Gradual improvement of understanding of value-driver relationships • Information repository to decide what facts to add in your replicable reference external communications	• Value-driver based analysis still in its infancy • Noise from other effects

CHAPTER 9

Design/Strategy

The Vacuum in the Majority of Digital Transformations

Before we start to dive into the importance of strategy (aka the design element of our framework) for your digital transformation, let us revisit a story from our childhood days, from the brothers Grimm. It very nicely describes the gist of everything that follows in the remainder of this chapter.

The Hare and the Hedgehog

The hedgehog was still quite near his home and was just rounding the blackthorn shrub on his way to the turnip field, when he observed the hare, who was on his way to visit his cabbages. The hedgehog bade the hare a friendly good morning. But the hare, who was frightfully haughty, did not return the hedgehog's greeting. Instead, he said contemptuously, "How do you happen to be here in the field so early in the morning?"

"I am taking a walk," said the hedgehog.

"A walk!" said the hare, with a mocking smile. "It seems to me that you might use your legs for a better purpose." This answer made the hedgehog furiously angry. His legs worked quite fine for his purposes, but they were a bit short and crooked, a fact that made him feel slightly defensive. So the hedgehog replied, "You seem to imagine that you can do more with your legs than I with mine."

"That is just what I do think," said the hare.

"That can be put to the test," said the hedgehog. "I'll bet that if we were to run a race, I would beat you."

"That is ridiculous! You with your crooked, stubby legs!" said the hare, "But for my part I am willing, if you have such a monstrous fancy for it. What shall we wager?"

"How about a golden coin and a bottle of brandy," said the hedgehog.

"Done," said the hare. They shook hands to seal the deal. "Well, we may as well do it right now, don't you think?"

"Nay," said the hedgehog, "there is no such great hurry! I haven't eaten yet, so I will go home first and have a little breakfast. In half-an-hour I will meet you back here."

The hedgehog went home, leaving the hare in the field. The hedgehog thought, "The hare relies on his long legs, but I will get the better of him. He may be wealthy, but he is a very silly fellow, and he shall pay for insulting me." So, when the hedgehog reached home, he said to his wife, "You must come out to the field with me."

"Why? What's happened?" said his wife.

"I have made a wager with the hare, for a golden coin and a bottle of brandy. I bet that I could beat him in a race, and I want you to see it."

"Good heavens," cried the hedgehog's wife, "have you lost your mind?"

"Don't worry," the hedgehog told her. "I have a plan, but I need your help." The pair walked out to the end of the field nearest to their house. The hedgehog told his wife, "We'll use this long field for our race-course, with each of us running in separate furrows. We'll start at the far end of the field and the finish line will be right here." The hedgehog pointed to the end of the furrows just inches from their feet. "All I need you to do is to lie down at the bottom of this furrow, here at the finish line. When the hare arrives, shout, 'I am here already!'" The hedgehog and his wife looked and sounded just like one another, so he was certain this plan would work.

"Okay," said his wife, taking her place in the furrow. The hedgehog walked to the far end of the field, and the hare was waiting for him there. "Shall we start?" said the hare.

"Certainly," said the hedgehog. Each placed himself in his own furrow. The hare counted, "Once, twice, thrice, and away!" and went off like a whirlwind down the field. The hedgehog, however, only ran about three paces, and then he stooped down in the furrow, and stayed quietly where he was. When the hare arrived in at the finish line, the hedgehog's wife met him with the cry, "I am here already!" The hare was shocked. The hare, however, thought to himself, "That hedgehog must have cheated somehow," and cried, "Let's have a do-over! This time, we'll race back to where we started." And once more he went off like the wind in a storm, so that he seemed to fly. The hedgehog's wife stayed quietly in her place, so when the hare reached far end of the field, the hedgehog himself cried out to him, "I am here already." The hare, however, quite beside himself with anger, cried, "It must be run again, we must do it again."

"All right," answered the hedgehog, "for my part we'll run as often as you choose." So, the hare ran 73 times more, and the hedgehog always held out against him, and every time the hare reached either end of the field, either the hedgehog or his wife said, "I am here already."

At the 74th time, however, the hare could no longer reach the end. In the middle of the field, he fell to the ground, blood streamed out of his mouth, and he lay dead on the spot. But the hedgehog took the golden coin, the bottle of brandy, and called his wife out of the furrow. They both went home together in great delight, and if they are not dead, they are living there still.

There are many morals you could get out of this story. But for digital transformation it is very simple and clear. Do not be the hare! You may think you will grow longer legs and get faster by transforming your organization (despite your various degrees of mounting panic), giving everything you have, leveraging on all the costly digital transformation payday elements (catalysts, reactants, reaction mechanisms, and their representation in products) described earlier. But when you have a competitor like the hedgehog, you will drop dead sooner or later having gained nothing. Instead, be the hedgehog, and find your hedgehog wife! Find something unique that your competitor did not see coming, which will give you the ultimate sustainable advantage in the game you have chosen to play.

After all, you are in this game to win. For me, *design/strategy* is the most important success factor in digital transformation practice. Therefore, it should come as no surprise that strategy ended up prominently in our applied digital transformation payday framework (see Figure 3.2). Nevertheless, I decided to explain it in this part of the book as I believe it makes much more sense to first understand in detail the key payday accelerators and decelerators before exploring the other digital transformation elements. That way you know what must and can be leveraged in your design/strategy to configure an end-to-end path toward your desired target.

My insistence that it is critical to develop a strategy and make decisive choices might sound like theoretical, old-school business history. Disciples of high-speed, disruptive market dynamics driven by customer needs—or those who believe that swift, agile execution is the ultimate savior of any business—might think my approach is stodgy or outdated. But believe me, given what I have seen in many digital transformations, it is not. Quite the opposite: To my knowledge no one has reasonably proven in practice that a religious focus on execution alone (as embraced by some wrong agile interpretations) can replace good strategy work. Only through strategy can you reach the goal of developing a lasting competitive advantage rather than some incremental improvements that cannot ever set you apart. Benchmarking or simply going with what your customers ask for at high speed, while relying on the same digital experience and cloud platforms as everyone else, will not be sufficient for the sustainable success of new business models. It will work even less for large corporations with their many frontier, adjacent, and core reactants. A clearly defined strategy and target state is an integral part of embarking on any meaningful digital transformation, this fundamental belief is not as broadly accepted as one would hope for. Consequently, in many digital transformation programs we often see what I consider to be a very dangerous vacuum which has proven to substantially increase the likelihood of a transformation implosion. (The failure rates needs to come from somewhere). Experience has shown that even the best-running digital transformation machine needs to know where it should be going. Without a clear strategy to win in the marketplace—that is, the search for something that will make your firm unique vis-a-vis your competitors, as the foundation of your most certainly long and very painful digital transformation journey—your digital transformation is doomed to fail before you even start. Without strategy, in the best-case scenario, you will end up executing successfully and then find out that all the money you invested in the process has gained you nothing vis-à-vis your competitors. You, as the hare, will meet your hedgehog nightmare.

Why Good Strategy Is Now More Important Than Ever

But do not worry. This is not another strategy book. So do not expect a lengthy elaboration to make my point of what I believe is now the final best way and coolest new framework for developing winning strategies. There have been just too many (Mintzberg, Ahlstrand, and Lampel 1998) and they keep coming. Nevertheless, I obviously have my personal favorite. Based on trying numerous different concepts over more than 20 years of strategy consulting, I have decided to stick to the one that worked best for me and, more importantly, for my clients: the "Play to Win" approach (Lafley and Martin 2013; Martin 2021). But no matter how you decide to develop your own strategy in the end, just do it! I am utterly convinced that the single most likely root cause of destruction for any digital transformation success is not having one at all, which is a (very stupid) choice by itself; "Strategy is not what you say but what you do." (Martin 2021). But beware! Having a strategy is not the same as a "strategic plan" with endless presentation pages developed by consultants or internal strategists sharing their sophisticated analysis of the past (what is rightly called "boiling the ocean"). This will not help you with forecasting and road-mapping the future. A true strategy is:

- A creative exercise leading a clear, target-state ambition and
- a fitting configuration of your integrated business and operating model that
- cannot easily be copied by competitors of whatever current or future flavor.

Therefore, such a winning strategy should define a target-state that will give you (at least for some time that matters) a sustainable advantage for winning in your chosen marketplace.

This also implies that this strategy cannot be a purely one-off exercise. The strategy you need to have in place for your digital transformation to succeed and secure its payday has nothing to do with the misconceptions of waterfall-style strategic planning. Instead, we are talking about a constantly repeated (this is where I agree with the agile disciples) process of hypothesizing a potential winning target-state together with integrated choices regarding business model and operating model. Through this process, you learn what works and what does not, and then start anew. To be even more precise: You do not need an extra for your digital transformation to succeed, it is the other way round. You need to design your digital transformation so that it fits your overall strategy choices for winning.

The Sense and Nonsense of Digital Strategy

Therefore, it should come as no surprise that the often-heard request or recommendation for developing a "digital strategy" makes the hair on the back of my neck bristle. *Digital strategy* is utter nonsense. For me there is no such thing. As explained in earlier chapters, digital has many impacts on accelerating or decelerating your transformation payday. Obviously, you might develop a winning strategy where digital plays a prominent role because exactly these accelerators (after swallowing the impact of decelerators) give you some unique advantage others in the market cannot copy easily. So do not feel pressured by anyone: You might need to revise your strategy for the digital age, but you certainly do not need and should not have a separate digital strategy. It becomes even worse when firms try to develop multiple "strategies" for each and every catalyst we have discussed before—cloud strategy, AI strategy, RPA strategy, cognitive strategy, and so many more.

From my perspective you can call these concepts whatever you like, but you are only referring to an implementation approach and plan, which can and should support what you are hypothesizing to achieve with your strategic choices. In most cases I know, that requires several of these concepts working together in combination to build something unique and not easily replicated.

CHAPTER 10

End-to-End

Our Framework in Action

To better understand how all elements of our framework fit together in practice, let me briefly sketch two recent digital transformation examples from real clients. They are heavily sanitized for obvious compliance reasons, but you should get the idea.

Example 1: Adjacent/Frontier Business Transformation

The first client is a publicly traded technology services provider, among the top five companies in its current markets and an aggressive challenger to the biggest two players operating in the space. Its existing services and customer base were growing at a moderate pace and at low but acceptable margins, but given the overall digital transformation hype, it was lacking the fantasy of something new and unique. To initiate the transition to a more digital-driven company, the CEO relied on the bet that there would soon be a new B2B2C market segment in their sector. He envisioned a **design/strategy** aiming at swiftly establishing a new IoT (Internet of Things) proposition based on building a unique cloud and experience technology platform of platforms, combining leading-edge analytics of existing sanitized customer data built on unique cyber-/data-security technologies as **catalysts/drivers**, a soon-to-be developed IoT service toolbox environment, and an open-access digital-experience technology model. This open-access model would allow third parties (the Bs in B2B2C) to develop their own services based on the data and services on the platform at lower cost and sell them over their own channels and customers (the Cs in B2B2C) for a fair service fee at good margins. It was an assumed win-win for everyone involved based on the synergies generated. This kind of transformation was considered too disruptive to be driven out of the core business, so a new adjacent/frontier **reactant/scope** company in a different, "cooler" city than the headquarters location was set up. It was staffed with only a handful of key managers from the core, plus a completely new team built from scratch. The result was a different employer brand, a more digital culture, and all the other prerequisites to attract the digital workforce and fulfill their needs, including new ways of working and a newly designed office space. To further increase momentum as an accelerator, an IT services company was onboarded in a build and transfer model with significant success-driven enumeration. Another firm was tasked to search the start-up ecosystem for potential M&A (mergers and acquisitions) targets to further boost the business platform with existing customer relationships and analytics and new cognitive technology. As the platform did not yet exist, a large-scale hybrid **reaction mechanism/process** was established to develop it, with a clear waterfall roadmap to measure the success of the company and the consultants and every element of agile development processes you can imagine.

The planned results of becoming the leader in a new digital market segment were happily communicated in the form of replicable references as **products/outcomes** to all stakeholders, even adding some financial key performance indicators (KPIs) to the perspective.

Before you ask: No, there was no digital transformation payday ever. The opposite happened and the whole model was closed less than two years after starting. The adjacent company is now history and all managers in charge had to leave the parent company soon thereafter. While the idea as such was certainly worth exploring, the decelerators pulling down the transformation were unfortunately heavily underestimated and all accelerators succumbed quicker than anyone could have imagined. The managers in the core business always viewed this new model with suspicion (there were legendary meetings with 30 advisors and two poor start-up founders in the M&A processes). It became clear that the timelines and revenue expectation defined in the waterfall approach would not hold, attracted partners for the platform business model where too small to ever achieve sufficient scale, IoT devices were in scarce supply. When the workforce onboarded did not comply with the requirements of central HR, core business managers started questioning the overall idea the moment the CEO left the company in the middle of the transformation. The business had no further chance of survival. Add one more to the failure stories.

But all this was not for nothing. The learning curve was steep. Nowadays new start-up business models are first tested in an incubator as part of a larger portfolio of ideas with one key filter criteria: the core business must commit to being the first and a significant internal customer of what these adjacent/frontier businesses are offering. Once the model has stabilized, it is fully integrated into the core business until the next one comes along and so on.

For a simple overview, Table 10.1 summarizes all elements and highlights some selected accelerators and decelerators from hindsight.

TABLE 10.1 Example elements of adjacent/frontier business transformation.

Digital Transformation Payday Framework Element	Description	Accelerators	Decelerators
Design/strategy	• Cloud- and digital experience technology platform of platforms for IoT B2B2C services • Hard-to-replicate assets in place (customer data treasury) • First-mover advantage, no competitor yet in the space		
Supply-side catalysts/drivers	• Innovative cyber/data security and analytics technologies • Out-of-the box functionalities in standard digital experience/ cloud platforms	• Built-up new platform at higher speed and lower cost	• Exploding cyber/data security concept and development cost to satisfy regulator that was not readily available out-of-the-box • Integration cost for existing customer data platform underestimated
Demand-side catalysts/drivers	• New B2B customer segment aiming at monetizing lower-margin IoT business models at low based development cost	• New unaddressed B2B2C market segment with potential of high growth	• Explosion of development times and cost due to business requirement surge from this segment • Underestimated shortage of skilled developers, leading to requirement to onboard external consultants for build and transfer at a substantial budget

(continued)

TABLE 10.1 (continued)

Digital Transformation Payday Framework Element	Description	Accelerators	Decelerators
Reactants/scope	• Newly established adjacent business unit in a location away from HQ	• Quick setup • Lean structures • Access to new talent pools in different city	• Culture clash between adjacent and core • Reintegration plan and cost not factored in • Key managerial talent lost for the core after failure of business model, they left the company
Reaction mechanisms/processes	• Waterfall roadmap, agile software development	• Higher speed feature development cycles • Early launch of minimum viable product to generate first mover advantage	• Friction between waterfall roadmap and gradual platform improvements • Substantial governance cost and efforts to align on progress with core business
Products/outcomes	• A revenue forecast for the new entity		

Example 2: Core Business Transformation

The second client is a large capital-market-listed infrastructure services company operating on a 15-year-old, outdated, on-premises legacy IT-system landscape across all its front- and backend processes. After a long period of repeated cost-cutting programs with hardly any revamp of systems, leading to among other things, a severe deterioration of customer experience, KPIs management decided to launch a liberational transformation to massively refocus the company on customer centricity.

To do so, a chief commercial officer (CCO), with the agreement of all other executives, envisioned a **design/strategy** aimed at providing an industry-unique digital experience and cloud-technology enabled customer experience based on never-before-seen simplicity of offerings and service processes at fair and transparent prices, thereby dismantling many of the industry's traditional complex tariff schemes. The redeeming idea to make it happen was to select one single innovative digital-experience-technology vendor as a **catalyst/ driver** to implement its mostly out-of-the-box, preintegrated industry solution to achieve the required large improvements in one big leap, despite the worry and lack of clarity about how much of the current business models could be covered by this solution and how much would have to be customized. The benchmark ambition was not set by any competitors in the industry, but by the state-of-the-art processes implemented by hyperscalers, the ones customers mentioned in market research as the real companies to compare.

Consequently, the **reactant/scope** of transforming toward the required target-state configuration was nothing more and nothing less than the core business, across all customer segments and all services. This was unanimously decided with a full understanding of the long and painful journey it would entail.

Due to the various questions about the vendor's ability to deliver, management decided to follow a very classical plan-based waterfall **reaction mechanism**, starting with an extensive multidimensional process design and program-planning phase before letting the vendor embark on its implementation process.

As in our first example, the waterfall-planned endgame of becoming the digital experience leader in the market was aggressively communicated in the form of replicable references as **products/outcomes** to all stakeholders, together with a "fit-to-purpose" business case developed by an external consulting team.

As you might have already guessed, the program failed catastrophically. The dreamed-about digital transformation payday never came. After a lengthy multidimensional and immensely complex process design and planning phase to ensure business as usual, it became clear to everyone involved that instead of an 80–90% coverage of requirements by the vendor's out-of-the-box functionalities, required customizing share was trending toward more than 70%, with obvious drastic implications for development times and, more importantly, budgets and a massive demand of skilled workers or externals. The very much wish-driven business case went south before the implementation even started. After several highly dynamic board offsites, management finally decided to pull the reigns. The vendor was fired, many key people on all levels left the company soon thereafter. Communication to shareholders about the program was silently phased out.

But again, all these investments wasted were not for nothing. After a year of regrouping a new program with the same objectives was started, but the setup was very different. Instead of a single vendor, a best-of-breed portfolio of leading cloud-based solutions was selected. The business case was set up as a mostly self-financing vehicle by onboarding a large system integrator incentivized on replacing the legacy systems at a 30% cost improvement, and the whole development process was established as hybrid, with rough roadmap milestones but a completely agile way of working for all integrated business and IT teams. While there have been many ups and downs so far, things are currently progressing somewhat better than they did in the first instance. Still, digital transformation payday is years away and it will require a lot of resilience to make it happen in the good and bad times of the company still ahead.

For a simple overview, Table 10.2 summarizes all elements and highlights the major accelerators and decelerators from hindsight.

TABLE 10.2   **Examples of elements of core business transformation.**

Digital Transformation Payday Framework Element	Description	Accelerators	Decelerators
Design/strategy	Digital-enabled, hyperscaler-like customer experience First-mover and unique to the current industry segment		
Supply-side catalysts/ drivers	New out-of-the-box cloud- and digital-experience technology platforms	Preintegrated cloud-based industry solutions	Unexpected gap between the advertised out-of-the-box functionality and business-critical requirements
Demand-side catalysts/ drivers	Hyperscalers as the cross-industry experience benchmark	End-customer frustration with all traditional industry players	High salary demands of very scarce skilled workers leading to the need to onboard more externals than planned
Reactants/ scope	Core business, across all customer segments and all services	Largest lever addressable in one leap, more than 90% of overall revenues affected	Extremely complex requirements needed to ensure BAU (business as usual) for all segments
Reaction mechanisms/ processes	Classical plan-based waterfall	Clear roadmap with measurable milestones	Late discovery of plan-deviations and ultimate cancellation of the overall program
Products/ outcomes	Operational cost savings and revenue-upside business case		

CHAPTER 11

"Smaller" Firm Excursus

David versus Goliath?

A s explained already earlier, this book is mostly based on research and experience from large (capital-market-listed) firms. But some readers will rightly ask, "What does this all mean for 'smaller' firms? Can one assume that the accelerators and decelerators follow the same rules when you are a David-like company fighting for winning in a market full of Goliaths?"

Yes and no. In general, everything I have said about all elements in our digital transformation framework I also consider relevant for nonlisted smaller firms, but there are some specificities. Some elements make your life easier and allow for additional acceleration; others will slow you down in your quest for your digital transformation payday. There is plenty of literature on scaling successful start-ups, but I guess one could write a whole new book on the interesting issue of digital transformation in small to midcap firms or even family-owned businesses. As I have decided to focus on large firms in this book, Table 11.1 briefly summarizes some interesting initial overarching implications for the key elements when you are a smaller fish in a big pond. I can only encourage interested researchers to spend more time on this issue.

Throughout Part II, you have encountered detailed discussion of the framework's elements. Using this table, you can refer back to the chapter to better understand how to most effectively apply the elements in a small-business context.

TABLE 11.1 Some general implications from a small firm perspective.

Framework Element	Subcategory	Smaller Business-Specific Accelerators	Smaller Business-Specific Decelerators
Design/ strategy	–	Less complex portfolio of strategic choices could allow for more timely winning strategy development	Every market impact relevant choice becomes a bigger "bet" with therefore higher risks for the overall firm in case of failure
Catalysts/ drivers	Technology capabilities	More opportunities for experimenting with lower-cost tech due to potentially less strict requirements on scalability, security, etc.	Overall platform cost (especially for implementation) not necessarily proportionally lower with size, especially when implementing market leading vendors Cross-technology strategic benefit amplification harder to achieve when firms can only afford a limited number of parallel implementations
	Workforce skills	Smaller-scale skill legacy, allows for faster workforce transformation	Existing skillset potentially even more traditional, fewer pockets of innovation from where to start the spark of workforce transformation
	Funding vehicles	Can experiment easier with smaller-scale innovative financing options	No access to many bigger options, especially when capital market related
	Customer needs	Less complexity, quicker time-to-market for improved customer journeys	Customer journey expectations are driven by what the market leaders do, little patience for the followers
	Workforce expectations	Less legacy, easier to attune culture and processes for the requirements of the new type of digital workforce	Many: For example, employer brand recognition, requiring workforce to move to B-locations with less attractivity outside of the major digital hubs, required salary markups

Framework Element	Subcategory	Smaller Business-Specific Accelerators	Smaller Business-Specific Decelerators
	Spending patterns	New pockets of value, previously not addressable by smaller firms	Disproportional investment required to achieve minimum required share of attention
	Blurring industry boundaries	Even for smaller companies, opens new markets with previously prohibitive cost of entry	Possibility of entry of larger firms at scale into previously better protected niche businesses
Reactants/ scope	Core	Potentially less complex and costly to transform legacy business	Comparatively requires bigger "bets" with higher risks for the overall firm in case of failure
	Adjacent/ Frontier	Less likely reintegration issues in case of success	Can quickly overshadow the core, if going wrong
Reaction mechanisms/ process	Waterfall	Can be a more natural fit to some smaller firms' traditional governance structures	Can increase decision bottlenecks in some smaller firms' traditional governance structures
	Agile/Hybrid	Less complex legacy structures potentially easier to transform into agile ways of working	Can be a diametral culture show for some smaller firms' traditional governance structures
Products/ outcomes	Subjective	Easier-to-analyze scale and complexity allow a better understanding of interdependencies in the maturity model chosen	The hard to escape enticement for oversimplification, leading to wrong maturity conclusions
	Objective	Less complexity should simplify achieving transparency at lower platform implementation cost (e.g., for NPS measurement or integrated cloud-based planning)	Could require new measurement platforms to be implemented sharing the problems of all technology capabilities (see earlier in the table)

PART III

Three Predictors in Your Business That Influence Your Digital Transformation Payday

CHAPTER 12

Some Groundwork for Prediction

Does Digital Transformation Cause Paydays? Well, It Depends . . .

I can already hear some readers saying what I would say in their shoes: *"Only solutions, not even more complexity please!"* And you are definitely right. From a strategy consultant one would expect a much clearer message on this decisive question. In my first draft it said: "Yes, but . . ." However, in Germany we have the nice saying that *"everything you say before the 'but' is a lie,"* so I decided not to make a change.

Nevertheless, "Well, it depends . . ." is certainly also not what you deserve to hear.

But it was written like this on purpose to provoke curiosity, deeper thinking, and an acknowledgment of the question's complexity. While this book is not meant to be a bulletproof, triple-peer-reviewed science publication (using statistics as *the art of never having to say you're wrong*), I am also a business school professor after all. *Digital Transformation Payday* has the ambition of leaving behind the purely opinion-driven fact-free world of digital transformation practice. Therefore, before we dig deeper into some potential predictors in your business that can influence your digital transformation payday, we need to develop a better understanding of the general relationship between digital transformation and market capitalization or earnings, respectively.

A scientific publication or not, any careful claim of indications for causality—and I am sure that is what you as a reader have been looking for—cannot be made seriously without careful analysis. Remember: We are looking for signals that an improvement of digital transformation replicable references (DIGITALPROXY) leads to a change in our chosen value variables like market capitalization (MARKETCAP) and future earnings (ROA3Y), ceteris paribus (other things being equal). This is especially so because we do not have the gold standard of causal analysis (experiments) in our hands, so we must instead purely rely on very noisy observational market data. This is a tedious and almost impossible task. Why noisy? In our complex business world there are uncountable effects influencing how a firm's value evolves. No matter how important any digital transformation program may be, it is never alone. Not only that, but you also must live with the annoying reality that your competitors are also transforming while you are. You are part of a game that will be decided by the best winning strategy being executed rigorously.

I can imagine that some of you were excited to directly transfer some sort of sophisticated quantitative assessment of all the wonderful accelerators and decelerators we discussed before into causal market value implications. That was what the first test readers of the initial sections of this book told me and this is how many typical "pragmatic" consultants would do it. I am very sorry to disappoint you, but this is not how it works.

Remember what we already concluded in Chapter 3:

The empirical parts of this book focus on the firm and aggregation levels. Only here, digital transformation can be externally observed without insider knowledge and only here capital market value development analysis can happen, given the irrelevance of the individual and nontransparency and incompleteness of project-level information from an external perspective. These nevertheless play an implicit role when we discuss the framework . . . and the inherent digital transformation payday accelerators and decelerators in each element . . . References to macro industry-/market-level perspectives must obviously be included on several occasions to reflect the need for an overall view on the digital transformation process. This is true especially when discussing the drivers or trigger points for transformation . . . and as a key element of the applied statistical models . . .

This means that whatever happens in between all our carefully designed elements with their accelerators and decelerators cannot be used as any basis for our analysis, unless—and that is the key task you have as an executive and manager—it materializes, based on a successful end-to-end execution of our overall framework, as visible replicable references (our variable DIGITAL-PROXY), operational KPIs, or even financial KPIs for the outside world.

Therefore, to better understand digital transformation's impact on market capitalization (measured in the variable MARKETCAP) and future earnings (represented by the variable ROA3Y) for listed large corporates, substantial effort was invested into understanding and selecting the best fitting statistical models and more certainty-generating robustness tests based on the scientific approach behind this book (see Appendix F). But which model to choose from this endless portfolio of options? Fortunately, we have already decided beforehand which variables we need to incorporate, based on at least some agreement in scientific literature and empirical research of what could be adequate: the "Ohlson econometric model" for our market capitalization analysis and a mixed fundamental econometric model for future earnings (see Chapter 3).

But regrettably, these econometric models do not tell us anything about which statistical model to use as none of them provides any clear guidance on the characteristics of their variables and the character of their interaction. Are they all linear in their relationships to the dependent variables MARKETCAP and ROA3Y, or are there any other functional forms relevant? Well, pick your guess for real-life. The answer for our business environment is as simple as the implementation implications are complex. They are nonparametric. To refresh your memory, nonparametric models have a distribution with no predefined functional form. (By contrast, linear models are based on a linear function.) This does not mean that any linear (regression) analysis is totally useless, but we need to verify whether any significant initial findings from these linear analyses also hold when we introduce nonparametric elements at least for DIGITALPROXY via semiparametric models or even become fully nonparametric for all variables by leveraging some recent innovations in machine learning.

Before we go down that route, we first start with clearly confirming the existence of fixed time and industry effects in line with Torres-Reyna's (2020) recommendations. This means that we need to factor in forces outside of our variables specific to an analyzed sector and a given period. As explained in Chapter 3 we mostly follow Muhanna and Stoel (2010) and build fixed industry/time effect models for the dependent variables MARKETCAP (1) and ROA3Y (5) as our "workhorse" solutions. To improve their statistical properties, both dependent variables were log-transformed (logMARKETCAP and logROA3Y). Finally, we supplemented our analysis with two corresponding fixed firm/time models ((2) and (6)), where fixed effects are reflected not on sector but on firm level.

To echo the earlier discussed need to check for nonparametric characteristics, exclusively for logMARKETCAP two further explorative and experimental approaches were added. Model (3) built on a very recent update from STATA (STATA 2020) and accommodated the hypothesis that at least DIGITALPROXY does not follow a fully linear function, with all other variables still linear. It therefore fitted a nonparametric function for DIGITALPROXY. Doing this does not make that much of a difference in the lower ranges of DIGITALPROXY, where we also find most of our observations as can be seen in the corresponding histogram of densities (Figure 12.1). For the higher ranges however, an overall semiparametric model resulted in visibly smaller confidence intervals (CIs), and therefore a comparably better fit to the real data (Figure 12.2). So nonparametric characteristics seem to matter! We cannot neglect them. Model (4) therefore went a few steps further and

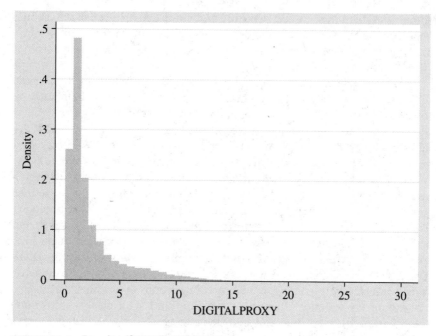

FIGURE 12.1 Density of DIGITALPROXY.

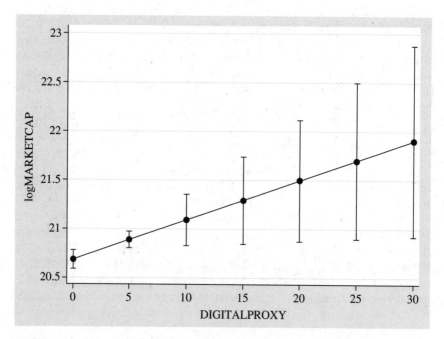

FIGURE 12.2 Linear logMARKETCAP predictive margins with 95% CI.

implemented a generalized, nonparametric orthoforest-based causal estimate for all variables.

> *"[Forests] . . . estimate very flexible non-linear models of the heterogeneous treatment effect. Moreover, they are data-adaptive methods and adapt to low dimensional latent structures of the data generating process. Hence, they can perform well even with many features, even though they perform non-parametric estimation (which typically requires a small number of features compared to the number of samples). Finally, these methods use recent ideas in the literature so as to provide valid confidence intervals, despite being data-adaptive and non-parametric. Thus, you should use these methods if you have many features, you have no good idea how your effect heterogeneity looks like, and you want confidence intervals. . . . Orthogonal Random Forests [short in this book: orthoforest] are a combination of . . . forests and double machine learning that allow for controlling for a high-dimensional set of confounders, while at the same time estimating non-parametrically the heterogeneous treatment effect, on a lower dimensional set of variables. Moreover, the estimates are asymptotically normal and hence have theoretical properties that render bootstrap based confidence intervals asymptotically valid."* (Oprescu et al. 2019)

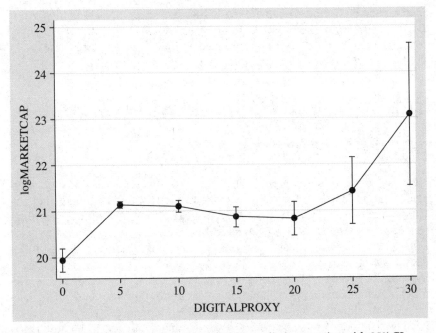

FIGURE 12.3   Nonparametric logMARKETCAP predictive margins with 95% CI.

The Rifle: Digital Transformation and Market Capitalization

Before we dig deeper into the depths of statistics for market capitalization, let me summarize the key results in simple language, so that you can skip the further details if you like: On average we have hit the mark for all companies and reports analyzed, we have found a statistically relevant relationship between DIGITALPROXY and market capitalization, both from a sector and from a single firm perspective. This does not mean that we have proven any directional causality, but it is certainly a starting point: There is at least some comfort in knowing that the higher a sector or firm scored on DIGITAL-PROXY, the more this could help to explain higher MARKETCAP, no matter which statistical model you apply.

Going deeper for market capitalization (logMARKETCAP), most results (see Table 12.1) confirmed our hopes for a statistically relevant relationship of DIGITALPROXY and of the majority of our other chosen explanation variables. As already expected, based on practical experience, and hopes, mean coefficients for DIGITALPROXY were >0 for these models. Our workhorse model demonstrated a satisfactory statistical significance (** $p < 0.05$). For such noisy observational market data the explanatory power is quite high, as measured by within R-squared (0.3072). The semiparametric and nonparametric models (3 and 4) show even higher expected impacts of increasing DIGITALPROXY on logMARKETCAP. All chosen financial and nonfinancial variables should be self-explanatory based on their naming, but in case of any doubt, you can revisit our explanation of the models chosen in Chapter 3.

The Shotgun: Digital Transformation and Future Earnings

How can results be summarized in simple words for future earnings? On average, for all industries/sectors and reports analyzed, there is a somewhat statistically relevant relationship between DIGITALPROXY, and future earnings as measured in logROA3Y. But statistical explanation power is very low, with a shotgun-like spread meaning that there are many other noncovered influences having a stronger impact. For the firm-level view, the impact was even more random, ranging from negative to positive, therefore statistically not significant, with even worse explanation power behind the results.

All this should at least be a warning that a higher sector or firm score on DIGITALPROXY cannot be assumed to automatically lead to anything that can be planned in terms of future earnings, neither a negative initial dip as

TABLE 12.1 Results of the main statistical models for logMARKETCAP.

Models	(1) Industry/Time Fixed Effects	(2) Firm/Time Fixed Effects	(3) Semiparametric (Digital Proxy)	(4) Nonparametric Orthoforest
Variables	**logMARKETCAP**			
DIGITAL PROXY	3.844e-02**	3.021e-02***	2.450e-01***	8.6177e-02#
	(1.822e-02)	(1.043e-02)	(2.589e-02)	
TOTAL EQUITY	4.422e-11***	1.858e-11***	omitted	
	(1.569e-11)	(3.226e-12)		
NET INCOME	1.02e-10**	3.716e-11***	1.04e-10***	
	(4.232e-11)	(6.593e-12)	(1.825e-11)	
AOCI	−9.68e-11**	3.980e-13	omitted	
	(4.469e-11)	(9.327e-12)		
PAYMENT OF DIVIDENDS	−1.74e-10*	−1.542e-11***	−1.82e-10***	
	(1.05e-10)	(5.285e-12)	(6.95e-11)	
DELTA EQUITY	−2.43e-10***	−1.891e-11***	−2.46e-10***	
	(5.32e-11)	(5.637e-12)	(4.765e-11)	

(continued)

119

TABLE 12.1 (*continued*)

Models	(1) Industry/Time Fixed Effects	(2) Firm/Time Fixed Effects	(3) Semiparametric (Digital Proxy)	(4) Nonparametric Orthoforest
REVENUE GROWTH	$-1.01\text{e-}05$	$7.86\text{e-}05^{***}$	$-2.77\text{e-}06$	
	$(3.29\text{e-}05)$	$(2.37\text{e-}05)$	$(4.00\text{e-}05)$	
L1.ROA	$2.994\text{e-}01^{**}$	$3.844\text{e-}02^{**}$	$2.999\text{e-}01^{**}$	
	$(1.246\text{e-}01)$	$(1.942\text{e-}02)$	$(1.190\text{e-}01)$	
NET DEBT	$9.628\text{e-}12^{**}$	$1.923\text{e-}12^{*}$	omitted	
	$(4.533\text{e-}12)$	$(1.119\text{e-}12)$		
INVESTED CAPITAL GROWTH	$9.26\text{e-}07^{***}$	$9.05\text{e-}08^{**}$	$9.38\text{e-}07^{***}$	
	$(6.96\text{e-}08)$	$(4.55\text{e-}08)$	$(1.06\text{e-}07)$	
BOOK TO MARKET	$-8.932\text{e-}04^{***}$	$-5.520\text{e-}04^{**}$	$-8.815\text{e-}04^{***}$	
	$(2.202\text{e-}04)$	$(2.261\text{e-}04)$	$(2.436\text{e-}04)$	
POLARITY	$2.105\text{e+}01^{***}$	$7.222\text{e+}00^{***}$	$2.065\text{e+}01^{***}$	
	$(3.668\text{e+}00)$	$(1.285\text{e+}00)$	$(1.083\text{e+}00)$	
SUBJECTIVITY	$-2.647\text{e+}01^{***}$	$-4.621\text{e+}00^{***}$	$-2.580\text{e+}01^{***}$	
	$(3.233\text{e+}00)$	$(1.220\text{e+}00)$	$(9.618\text{e-}01)$	

Models	(1) Industry/Time Fixed Effects	(2) Firm/Time Fixed Effects	(3) Semiparametric (Digital Proxy)	(4) Nonparametric Orthoforest
EARNINGS DATE DELTA	−7.464e-04***	−1.030e-04***	−7.461e-04***	
	(7.29e-05)	(3.07e-05)	(2.17e-05)	
R-squared	0.4256	0.9286		
Within R-squared	0.3072	0.0380		
Basic Fixed Effect	INDUSTRY	FIRM	INDUSTRY	INDUSTRY
Time Fixed Effect	YEAR	YEAR	YEAR	YEAR
Clustered Standard Error	INDUSTRY	FIRM	INDUSTRY	
Observations	22,176	22,176	22,176	22,176

Robust standard errors in parentheses.

*** $p < 0.01$

** $p < 0.05$

* $p < 0.1$ # Mean causal estimate: Stable to all key refutation tests. (see Appendix F for operationalization)

L1 = Lagged by one period

Empty = Not available or applicable

TABLE 12.2 Results of the main statistical models for logROA3Y.

Models	(5) Industry/Time Fixed Effects	(6) Firm/Time Fixed Effects
Variables	**logROA3Y**	
DIGITAL PROXY	2.696e-02***	1.620e-04
	(8.474e-03)	(1.295e-02)
TOTAL ASSETS	−1.125e-12**	−1.666e-12***
	(4.845e-13)	(5.617e-13)
NET INCOME	6.34e-11***	7.36e-11***
	(1.520e-11)	(1.238e-11)
NET INCOME GROWTH	3.850e-04*	5.167e-04***
	(2.042e-04)	(1.881e-04)
REVENUE GROWTH	−4.299e-04*	−5.769e-04***
	(2.279e-04)	(2.100e-04)
POLARITY	−1.952e+00	1.274e+00
	(1.370e+00)	(1.365e+00)
SUBJECTIVITY	−8.620e-01	−2.865e+00**
	(1.047e+00)	(1.188e+00)
EARNINGS DATE DELTA	−5.55e-05***	1.02e-06
	(2.05e-05)	(3.14e-05)
R-squared	0.4209	0.7688
Within R-squared	0.0264	0.0172
Basic Fixed Effect	INDUSTRY	FIRM
Time Fixed Effect	YEAR	YEAR
Clustered Standard Error	INDUSTRY	FIRM
Observations	15,241	15,067

Robust standard errors in parentheses.
*** $p < 0.01$
** $p < 0.05$
* $p < 0.1$
Empty = Not available or applicable

pessimists would think, nor the clearly demonstrated major profit contribution that has sometimes been claimed by other digital transformation books. You are certainly on safer ground relying on and communicating a digital transformation relationship to market value rather than to any earnings improvements.

In detail for future earnings (logROA3Y) only our industry model (5) shows the required statical significance (*** $p < 0.01$) for DIGITALPROXY. However, the statistical explanation power of both models (5) and (6) was not so convincing when you look at the within R-squareds (see Table 12.2) of 0.0264 and 0.0172, respectively. All the other chosen variables should be self-explanatory based on their naming, but in case there's any doubt, you can revisit our explanation of the models chosen in Chapter 3.

Some Tiny Steps Toward Causality

But do we really have to give up on our quest for causality? Even for market capitalization where statistically significant and relevant correlations are at least visible? Fortunately, not yet. There are some things we can do to get a clearer picture for our workhorse model and the nonparametric orthoforest estimates to establish some more confidence for any indicative causality claims.

Table 12.3 summarizes a simplified view on some causality requirements relevant for our nonexperimental data and the simple robustness tests conducted to soften any obstacles' impact. I have spared you most of the technicalities of how this has been done and what the theories are behind them and moved all that to the appendices. As you can see in our little overview, results are not 100% conclusive but still promising. Obviously, with our chosen mostly hands-on robustness tests, we cannot really prove any causality. Nevertheless, our analysis should at least make us much more confident that digital transformation and market capitalization are not only correlated in a statistically significant way but might even have a directional causal relationship from digital transformation to market capitalization. This makes it worthwhile to work on your digital transformation accelerators and decelerators to improve the value of your firm for its shareholders.

Potential Predictors: Recognize Your Starting Point

We now know that digital transformation, on average, seems to matter with some initial indication of causality for (market) value generation when looking back on our extensive dataset. But like the statistician who lays

TABLE 12.3 Robustness tests for causality indication.

Requirements for Causality/Robustness Tests	Correlation Analysis	Sensitivity Analysis (see Appendix F)	Instrumental Variables (see Appendix F)	Effect Direction (see Appendix F)
Association: An empirical or observed association between the independent and dependent variables	Positive and significant for logMARKETCAP Unclear and partially not significant for logROA3Y			
Time order/direction: The variation in the independent variable came before variation in the dependent variable—the cause must come before its presumed effect and the direction of effects is clear	Panel data used logMARKETCAP values from one theoretical second after DIGITAL-PROXY became public knowledge (the filing date) logROA3Y values end of year of DIGITALPROXY plus next two years on average			DIGITALPROXY >>> logMARKET-CAP Not checked for logROA3Y due to lack of significant correlations

Requirements for Causality/Robustness Tests	Correlation Analysis	Sensitivity Analysis (see Appendix F)	Instrumental Variables (see Appendix F)	Effect Direction (see Appendix F)
Nonspuriousness: The relationship between the two variables is not due to changes in a third variable, so what appears to be a direct connection is in fact not one.		Mixed for logMARKET-CAP Not checked for logROA3Y due to lack of significant correlation	Stable for logMARKET-CAP Not checked for logROA3Y due to lack of significant correlation	

Empty = Not available or applicable
>>> Arrows show direction of explanation power found

with his head in an oven and his feet in a deep freeze stating, "On the average, I feel comfortable," this does not help us more than generating a basic comfort that digital transformation is probably worthwhile.

Therefore, this book goes one substantial step further trying to help you better understand your odds of making your digital transformation payday, based on your starting position. In order to do so, the research looked even deeper into the dataset and tried to distill three potential predictor types for your digital transformation payday success.

To manage the complexity of potentially infinite digital transformation payday predictors they were, based on the extensive underlying research, clustered in three simple groups, which are explained in more detail in the following chapters:

1. Markets: Your sector/industry
2. Financials: Your balance sheet and P&L
3. Communications: Your communication approach

We all know that in a normal uncertain market environment, with many still-to-be-formed new strategies and external shocks, one cannot hope to predict the future from the past. And certainly not from models trying to describe what others have done. Nevertheless, looking into such elasticities for the first time ever allows us to gain some understanding of how other companies (the ones in our dataset) have fared with their digital transformation efforts depending on their context and behavior. This should give us some more confidence and knowledge on which we can develop our future digital transformation concept. You can get a better feeling about how others have fared in the past and reflect that in your strategy accordingly, be it by trying harder to manage the balance of your accelerators and decelerators, or by integrating additional risk buffers, or both.

CHAPTER 13

Predictor Markets

Different Industries Require Different Labors for Payback

You likely wonder: "Do all the learnings discussed before also apply to my sector or industry?" Therefore, this naturally leads to our first predictor group. For sure, practical experience and common sense would imply that the intensity and timing of digital transformation is very different by sector. The great news is that, with the unique data underlying this book, we do not have to rely on such gut feelings. While looking into history obviously does not imply applicability for the future, we can at least simply check the past!

It Only Goes Upward (for Digital Transformation Replicable References)

Before we do that, let us investigate the overall development over time. Despite significant noise from the data from 2011 to 2020 (Figure 13.1), a positive trend in the average share of replicable references as our proxy for digital transformation (DIGITALPROXY) can already be identified visually (Figure 13.2), reflecting the suspected increase after the surge of the digital transformation hype (with the highest scores in 2017).

Not All Sectors Are Created Equal

The promised deep dive from the sector angle, now looking at industry categories, confirms our expectation based on hard data. The obvious candidates for being more digital (e.g., computer hardware, computer software and services,

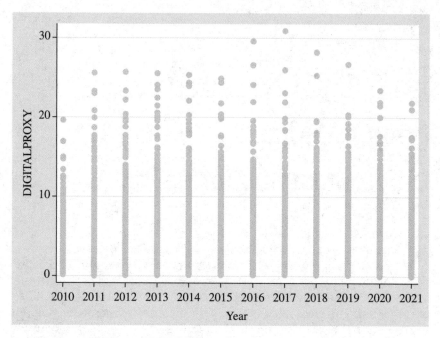

FIGURE 13.1 Noise in DIGITALPROXY over time.

consumer durables, internet, media, and telecommunications) show very different trends and higher means than more traditional industries (e.g., automotive, chemicals, drugs, utilities, insurance, and energy). See Figure 13.3.

Clearly, not all industries were embarking on the digital transformation journey in the same way, at the same time, and with the same verve. While this is interesting and confirmed our suspicions, it is obviously not giving us the answer we have been looking for yet: We want to know whether we can find different efforts required by industry category to achieve market capitalization impact, at least from a noncausal correlation perspective and based on our historical dataset.

On a more granular level of industry groups, it seems we can. By adding interactions of these industry groups to DIGITALPROXY into our main model (1) we can estimate correlation coefficients by each group interaction and check for the statistical robustness of each. Not all of them are robust, but you can still derive indicative findings for your specific situation from the following simplified overview (Table 13.1).

The clustering into positive impact/negative impact was achieved by taking the new DIGITALPROXY coefficient (which now becomes negative as a baseline) and adding the effect of the interaction. As you can clearly see, indeed not all industry groups are the same. For some, you can, with statistical significance, see a correlation to higher market capitalization for your digital

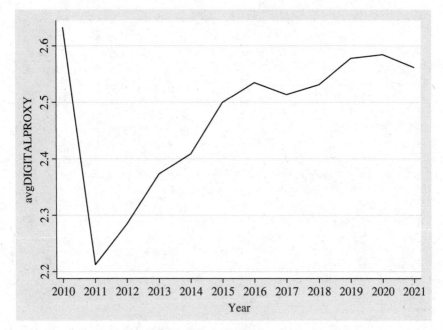

FIGURE 13.2   Development of average DIGITALPROXY over time.

transformation efforts in relation to all the other valuation model parameters than for others. There are expected winners, such as computer software and services, internet, and telecommunications, but also some unforeseen winners like metals and mining, and utilities, and some losers like energy and computer hardware. To spare you a long list of numbers here, I have moved detailed results into Appendix F. There you can also find the order of magnitude of correlations for your industry group. But mind you, as already mentioned they are not meant to forecast what will happen; they just give you some indication how for our dataset correlations have played out historically for others.

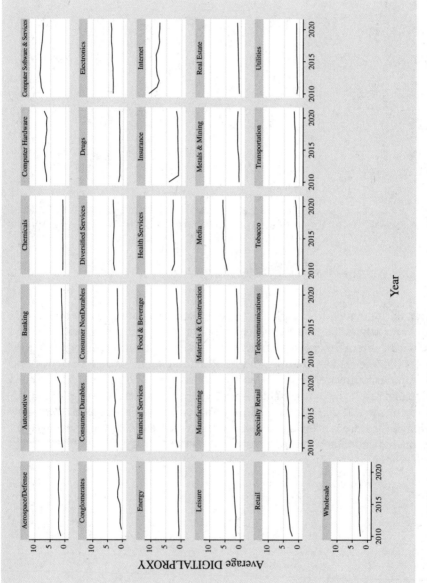

FIGURE 13.3 Differences in DIGITALPROXY by industry group.

TABLE 13.1   Interaction results by industry group.

Industry group total effects in interaction with DIGITALPROXY	Negative impact	Positive impact
Statistically significant	Aerospace and Defense (base)	Automotive
	Energy*	Banking
	Computer Hardware	Computer Software & Services
	Consumer Durables	Conglomerates
	Electronics	Consumer Nondurables
	Food & Beverage*	Diversified Services
	Insurance*	Drugs
		Financial Services
		Health Services
		Internet
		Leisure
		Manufacturing
		Materials & Construction
		Media
		Metals & Mining
		Real Estate
		Specialty Retail
		Telecommunications
		Transportation
		Utilities
		Wholesale
Statistically not significant	Chemicals*	Tobacco
	Retail*	

* = not significant slope difference from base (see Appendix F)

CHAPTER 14

Predictor Financials

Your P&L and Balance Sheet Context Indicate Your Ability to Achieve Your Payday

I ndustry groups are only one possible angle to cluster firms based on externally observable characteristics. Much more obvious to use are the financial key performance indicators (KPIs) of our analyzed firms. This should further enable us to better understand the context these firms are operating in. Fortunately, based on the advancements of econometric models, we can build on our orthoforest-based causal estimate model (Chapter 10) to run detailed elasticity analysis for DIGITALPROXY on logMARKETCAP. I again spare you the technical details here. (See some more explanation in Appendix F.) In simple words, this helps to answer the question about how the correlation impact (and indicative causality) of digital transformation efforts changed in our dataset depending on different observable financial KPIs and their characteristics (e.g., negative versus around zero versus high net income).

Table 14.1 shows the results, based on logMARKETCAP elasticity and a flag for the risk (*** = lowest risk) that the prediction is not as precise as one would wish (in terms of wideness of confidence interval), for all key operational KPIs. Given that such analysis is still in early stages, this should be further expanded in future research and will have to be carefully interpreted.

In any case, you can clearly see across all major variables, however, that firm context heavily influenced elasticities and the risks associated with embarking on a digital transformation as measured in confidence intervals around the mean effect. So, you can just pick some of your own firm's characteristics as a marker and at least find indications on how other companies from the dataset have fared with their digital transformation payday in the past.

There are three simple examples about how to read the table:

1. You are low on equity and produce negative net incomes: Firms with similar characteristics have demonstrated a *positive* correlation of their

TABLE 14.1 Orthoforest-based elasticity results for key variables.

Financial Variable/ Characteristic	Below Zero	Around Zero	Above Zero	Way Above Zero
TOTAL EQUITY	Positive**	Positive***	Positive***	Positive***
NET INCOME	Positive**	Positive**	Positive**	Positive**
AOCI	Positive/Negative way below/ below	Inconclusive**		
PAYMENT OF DIVIDENDS	Negative**/Positive** way below/ below	Positive**		
DELTA EQUITY	Inconclusive	Inconclusive	Inconclusive	Inconclusive
REVENUE GROWTH	Inconclusive	Inconclusive	Inconclusive	Inconclusive
L1ROA	Fluctuating*	Negative*	Positive*	Positive*
NET DEBT	Positive**	Positive**	Positive**	Negative**
INVESTED CAPITAL-GROWTH	Inconclusive	Inconclusive	Inconclusive	Inconclusive
BOOK TO MARKET	Negative	Positive*	Negative*	Negative*

*** = Tight confidence interval (very certain)
** = Medium confidence interval
* = Wide confidence interval
= Very wide confidence interval (very uncertain)
Empty = not applicable

digital transformation efforts (measured by DIGITALPROXY) to market capitalization with *medium uncertainty* (** CI).

2. You pay high dividends: Firms with similar characteristics have demonstrated a *negative* correlation of their digital transformation efforts (measured by DIGITALPROXY) to market capitalization with *medium uncertainty* (** CI).

3. Your book-to-market is negative: Firms with similar characteristics have demonstrated a *negative* correlation of their digital transformation efforts (measured by DIGITALPROXY) to market capitalization with *very high uncertainty* (no * CI).

Financials Matter: You Are What You Are

From a wide range of possible financials to analyze, this book has focused on our established underlying financial model (the Ohlson model, see Chapter 3 and Appendix E).

Remember, we used the following model to estimate market capitalization, with all financial variable names chosen to be self-explanatory. POLARITY and SUBJECTIVITY, as nonfinancial sentiment variables, will be discussed in the next chapter. DELTAEARNINGSDATE is a pure control variable and therefore not further assessed.

$$
\begin{aligned}
\text{MARKETCAP}_{jt} = {}& b_0 + b_1 \text{DIGITALPROXY}_{jt} + b_2 \text{TOTALEQUITY}_{jt} \\
& + b_3 \text{NETINCOME}_{jt} + b_4 \text{ACOI}_{jt} + b_5 \text{PAYMENTOFDIVIDENDS}_{jt} \\
& + b_6 \text{DELTAEQUITY}_{jt} + b_7 \text{REVENUEGROWTH}_{jt} + b_8 \text{ROA}_{jt-1} \\
& + b_9 \text{DELTAEARNINGSDATE}_{jt} + b_{10} \text{POLARITY}_{jt} + b_{11} \text{SUBJECTIVITY}_{jt} \\
& + b_{12} \text{BOOKTOMARKET}_{jt} + b_{13} \text{INVESTEDCAPITALGROWTH}_{jt} \\
& + \text{firm I industry} + \text{interactions} + \text{year} + e_{jt}
\end{aligned}
$$

For these variables we find:

- TOTALEQUITY: The elasticity of digital transformation efforts (DIGITALPROXY) depending on the book value of equity seems to be positive in all cases with medium certainty (medium confidence intervals) and declines with company size (Figure 14.1). At the high range, this effect almost becomes flat with high certainty (very tight confidence intervals). Apparently smaller companies (at least measured in equity) have some advantages here.

- NETINCOME: Net income demonstrates positive DIGITALPROXY elasticity values with medium uncertainty, with a steep increase once net income grows and then remaining positive with quite some fluctuation for mid to very high net incomes (Figure 14.2). It seems as if investors, at least for our dataset, have associated digital transformation with a future (growth) potential, even in case of very low or negative net income firms.

- AOCI: The accumulated other comprehensive income overall seems to have an uncertainty boosting elasticity effect for DIGITALPROXY. It becomes toward negative on digital transformation the higher it is (in this case shown by negative numbers) with wide confidence intervals. See Figure 14.3. So, beware if you are carrying loads of AOCI on your balance sheet.

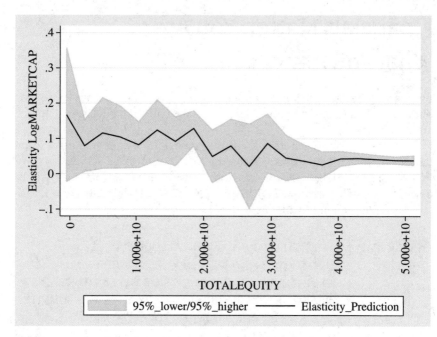

FIGURE 14.1 DIGITALPROXY elasticity TOTALEQUITY.

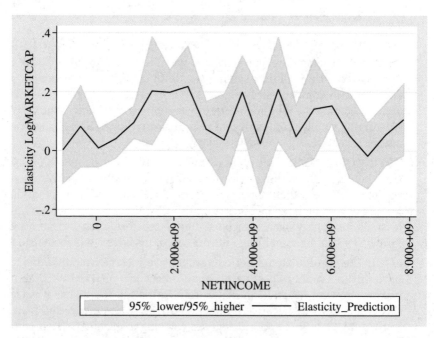

FIGURE 14.2 DIGITALPROXY elasticity NETINCOME.

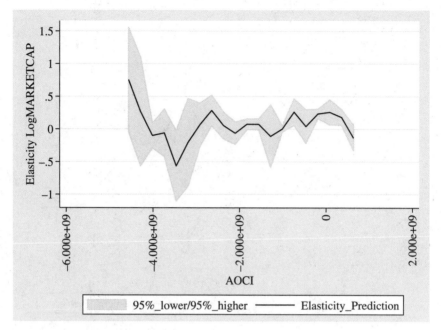

FIGURE 14.3 DIGITALPROXY elasticity AOCI.

- PAYMENTOFDIVIDENDS: Dividends paid support the intuitive notion that very-high-dividend-paying companies (in this case shown by negative numbers) will see less positive or even negative elasticities (Figure 14.4) for DIGITALPROXY. When higher dividends are paid out to shareholders, digital transformation, at least in our dataset, has more difficulty moving the digital transformation payback needle.

- DELTAEQUITY: The change in equity does not provide any relevant insights for our elasticity discussions. It can move the elasticity for DIGITALPROXY in both directions with very wide confidence intervals (Figure 14.5).

- L1ROA: If negative, the ROA of the previous period seriously hindered firms' ability to achieve positive elasticity impacts of digital transformation efforts (DIGITALPROXY), albeit with quite high uncertainty (wide confidence intervals) and quite some fluctuations. This situation only changed once positive ROAs were demonstrated, but uncertainty remained. See Figure 14.6.

- REVENUEGROWTH: The growth of revenue is unfortunately not giving us any further insights (Figure 14.7), as no clear patterns can be identified from our visual chart.

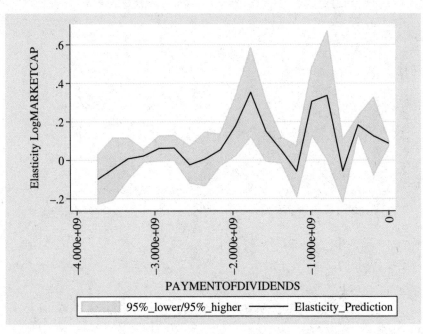

FIGURE 14.4 DIGITALPROXY elasticity PAYMENTOFDIVIDENDS.

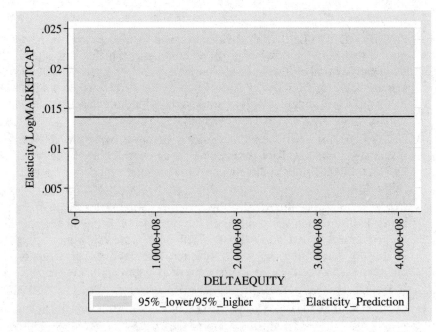

FIGURE 14.5 DIGITALPROXY elasticity DELTAEQUITY.

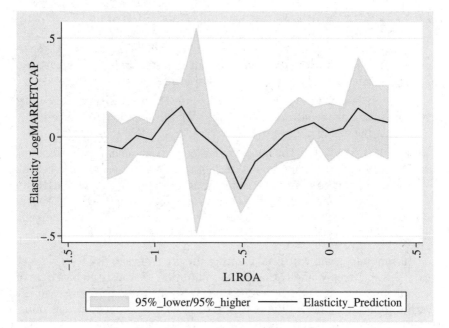

FIGURE 14.6 DIGITALPROXY elasticity lagged ROA (L1ROA).

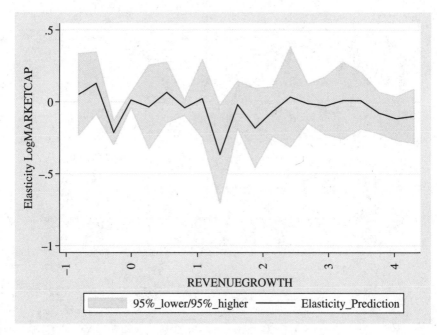

FIGURE 14.7 DIGITALPROXY elasticity REVENUEGROWTH.

- NETDEBT: The net debt elasticity impact for DIGITALPROXY remains stable positive with limited uncertainty when below/around zero, but steeply declines to negative with increasing net debt (Figure 14.8). Apparently, at least for the companies in our dataset, shareholders have not appreciated digital transformation efforts when there was a lot of debt.

- INVESTEDCAPITALGROWTH: The growth in the invested capital provided little insight because of very high fluctuation and uncertainty across the whole range. See Figure 14.9.

- BOOKTOMARKET: The book-to-market ratio is much more interesting, because, apparently for our dataset, companies with very low market values compared to their book values were in their mean, facing negative elasticities of their digital transformation efforts (DIGITALPROXY), but admittedly with very high uncertainty that the opposite could happen. See Figure 14.10.

In summary, even with all the uncertainties we have to live with because real life is just messy, and taking into account that obviously no past data of other companies guarantees you the same outcome for your future, one key learning point remains: For the companies analyzed, it was not only the industry they operated in that defined their ability to generate higher returns from

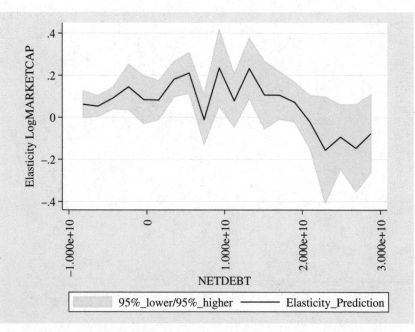

FIGURE 14.8 DIGITALPROXY elasticity NETDEBT.

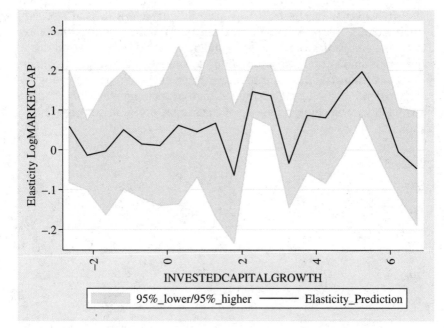

FIGURE 14.9   DIGITALPROXY elasticity INVESTEDCAPITALGROWTH.

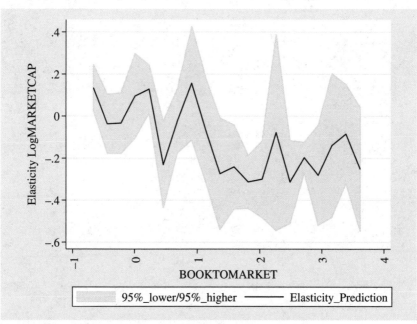

FIGURE 14.10   DIGITALPROXY elasticity BOOKTOMARKET.

digital transformation efforts; it was also the characteristics and strengths of their balance sheets. Shareholders seemed to rationally look at the whole context the companies were operating in and adjusted their views on digital transformation effort impact assumptions accordingly. Knowing who you are in terms of financials therefore becomes a key element when assessing the payday potential of your digital transformation program. And this is apparently not always playing out in favor of traditional, dividend- and asset-heavy companies.

CHAPTER 15

Predictor Communications

It Is About What You Sing and How You Sing to Influence Your Payday

This book and the underlying research emphasize analyzing what companies *say* about their digital transformation as the best yet available proxy for their true digital transformation status. For this we have heavily built on our proxy variable DIGITALPROXY, which measures what companies say about their digital transformation by counting what we have called replicable references in the annual reports. More information on how companies say what they say about their digital transformation (how they sing their digital transformation song) has so far only been in the form of sentiment controls for POLARITY and SUBJECTIVITY. They both showed positive and significant correlations to logMARKETCAP and logROA3Y.

But with all the analytical opportunities our nonparametric model offers, there are additional interesting questions we can ask:

- Which **sentiment** elasticities can be implied from the analysis on financial parameters based on our historical dataset?

- Can we go even further in somehow understanding the level of concreteness of what is being said, that is, what we call **quantification-level** elasticities?

For the sentiment elasticity case, we just rerun the same analysis (see Appendix F) as for our financials for POLARITY and SUBJECTIVITY.

For the quantification-level elasticity case, things are a little more complex. A more sophisticated dependency analysis was conducted for all digital transformation dictionary hits in the analyzed reports (see Appendix D). This investigation simply uses recent natural language processing advancements, a so-called dependency parser, to search for close syntax-based dependencies of the DIGITALPROXY hits to either date or monetary terms, and it brings

TABLE 15.1 Orthoforest-based communications elasticity impact.

Textual Variable/ Characteristic	Below Zero	Around Zero	Above Zero	Way Above Zero
POLARITY			Negative*	Positive**
SUBJECTIVITY			Inconclusive*	Negative**
D_CLOSE_M		Inconclusive*	Inconclusive*	Inconclusive*
D_CLOSE_D		Inconclusive*	Inconclusive*	Positive

*** = Tight confidence interval (very certain)
** = Medium confidence interval
* = Wide confidence interval
= Very wide confidence interval (very uncertain)
Empty = not applicable

these into a percentage relation to DIGITALPROXY. In other words, it measures: to what share were the DIGITALPROXYs found and further specified by dates (e.g., "We will implement Robotic Process Automation by January 2021") or values (e.g., "Our cloud migration program will lead to savings of 10m USD"). To simplify the scope of analysis for this book, out of the three conceptual measurement levels originally developed to describe concreteness in the original research, only the two measurements with a close syntax dependency relationship (less than five syntax arcs; see Appendix D) were used (D_CLOSE_D and D_CLOSE_M).

As these variables were not part of our main models for the book (mainly because in the basic main model, they were far from statistically significant with very high p-values >0.2), they were experimentally added in a separate step to form a new fully nonparametric model now including D_CLOSE_D and D_CLOSE_M. This model was then run for elasticities the same way as before.

Table 15.1 summarizes all results in the same format as in the previous chapter for financials. Detailed graphical outputs of the identified elasticities are discussed in more detail for both analysis categories thereafter.

Sentiment Elasticity Details: "Always Look at the Bright Side of Life" (Monty Python)

For our polarity measurement (POLARITY): If the general sentiment in your communication is not at the higher (bright side) end of the overall range

identified, even an intensified digital transformation communication potentially faces tough times to turn the tide of a more negative (dark side) sentiment context for your firm overall. Technically, this indication can be derived
from the fact that the mean expectation in our dataset, even for slightly positive polarities, is negative with medium confidence intervals (uncertainty)
and only turns positive and into tighter but not very tight confidence intervals
(less uncertainty) for higher polarities, as Figure 15.1 demonstrates.

For our subjectivity sentiment on the other hand (SUBJECTIVITY) we
see . . . that we do not see that much. Still, one could at least be tempted to
conclude that at the upper range of subjective report contexts, the positive
elasticity of pushing your digital transformation suddenly evaporates. This
would also fit very nicely to our business intuition that investors appreciate
objectivity and, at some breaking point of overly subjective communication,
lose their trust. Technically, as demonstrated in Figure 15.2, this indication
can be derived from the fact that the mean expectation, even for high subjec-
tivities (in this case above around 0.4), suddenly drops below zero and with
simultaneously tighter confidence intervals, signaling higher certainty. For
the subjectivity values below that range, the empirical data is not really allowing any clear conclusions.

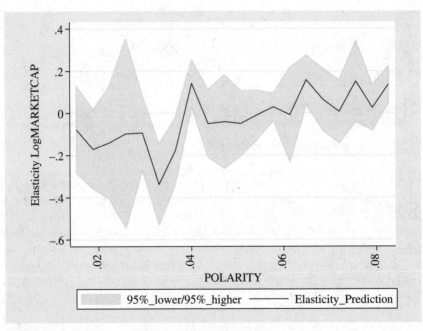

FIGURE 15.1   DIGITALPROXY elasticity POLARITY.

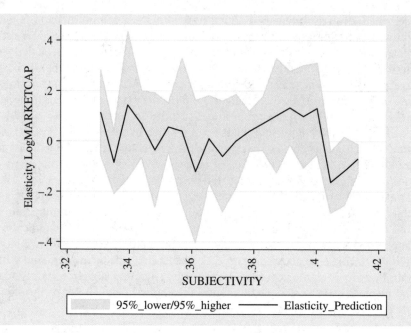

FIGURE 15.2 DIGITALPROXY elasticity SUBJECTIVITY.

Quantification-Level Elasticity Details: "Money for Nothing" (Dire Straits) or "One Moment In Time" (Whitney Houston)?

For both, the quantification-level measurement based on monetary terms (D_CLOSE_M) and the quantification-level measurement based on date terms (D_CLOSE_D), results are much less conclusive than what one would have hoped for. Not only does the mean effect of DIGITALPROXY drop below zero when we add these two, the elasticity curves produce very few insights we can further build on. While this should certainly be groundwork for further research, for the sake of our book, we can only conclude the following, as shown in Figures 15.3 and 15.4:

- At least based on our data, the share of monetary terms, other than what intuition would tell, does not give you any indication one way or the

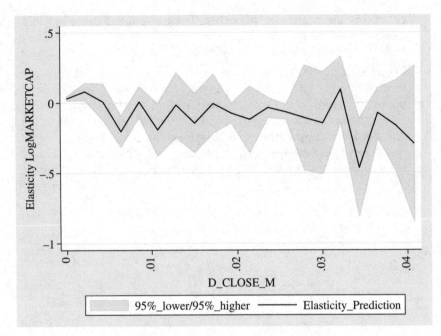

FIGURE 15.3 Quantification-level elasticities (monetary terms).

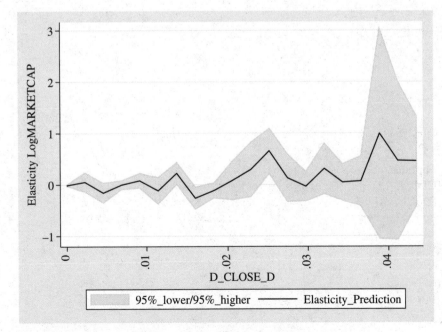

FIGURE 15.4 Quantification-level elasticities (date terms).

other that elasticities are fluctuating around plus/minus zero with a medium certainty based on the confidence intervals. So money can really be for nothing.

- The share of date (timing-related) terms in our data, other than what intuition would tell, does not give you any indication one way or the other that elasticities are fluctuating around plus/minus zero with a medium certainty based on the confidence intervals. Here, this at least changes for the upper range of date shares with a positive mean impact, but then with a substantial increase of uncertainty as exemplified by wide confidence intervals. So the moment in time is not totally futile.

CONCLUSION:
How to Keep Your Digital Transformation Paydays Coming

A Checklist to Speed Up Your Digital Transformation Payday Loop

For every complex problem there is an answer that is clear, simple, and wrong.

—H. L. Mencken (1920)

Why start the final part of this book with a statement written around 90 years before digital transformation even existed? Well, only to make it clear what this concluding chapter is and what it cannot and certainly will not be: a simple answer.

If you were hoping to jump straight to the easy-to-implement conclusion (or consuming some 10-minute summary via an AI-enabled app), I have to destroy your illusion right away: Despite what some managers, executives, shareholders, and even employees seem to believe, experience and the underlying deep research of this book show that just surfing on the hype and buzzwords, and copy–pasting a few concepts and recipes from new so-called best-practice examples that are churned out of the digital-transformation ecosystem on a daily basis, is the most likely common denominator of any failed digital transformation effort.

Instead, I strongly believe that for you to be able to succeed, you must pay utmost attention to all elements and all relevant accelerators and decelerators of your digital transformation journey in great detail and see them end to end across all interacting elements. There has never been such a thing as a value-generating business transformation without a clear winning strategy, endurance, patience, strong resilience to setbacks, and an eye for the detail and manifold complex interdependencies. As I hope I have demonstrated in this book, just adding digital as another component (or, even worse, as a "digital strategy" buzzword) does not change anything in this eternal truth. Instead, the age of digital often makes things even more challenging as many more potential and heavily interacting accelerators and decelerators for your payday are added to the equation. Therefore, while simplicity can and often should be an outcome of your digital transformation journey, if your strategy rightly aims for it as a key ingredient for winning, the transformation toward this goal usually makes things much more complex at first.

To make things even more challenging: There is no such thing as a single plannable digital transformation payday. Our framework shows that dependency flows in all directions for a good reason. Remember what we said in our framework description in Chapter 3:

> *All elements can and will influence each other in many directions.*
> *This means that not only a multitude of catalysts driving your trans-*
> *formation all at the same time can define which scope . . . is affected*
> *most, but also in return the transformation scope can define which*
> *catalysts (for example, digital technologies) have the highest impor-*
> *tance for the digital transformation process . . . The same is also true*
> *for the value backflow from digital transformation outcomes, which*
> *in return can influence all . . . elements before them, depending on the*
> *actual impact achieved . . . even though the described elements do not*
> *necessarily always create direct outcomes, they can and often will.*

So, the journey to your digital transformation payday in real-life is not a sequential process after all, but rather an interaction of complex multidirectional digital transformation payday loops. If you succeed in designing your digital transformation journey across all the necessary elements, and do it in a way so the accelerators ultimately dominate the decelerators, you should be able to speed up not only your next payday, but also build a virtuous cycle so that the paydays keep coming.

Even though there are no simple answers, there are still a few things you need to consider when embarking on a journey to your digital transformation payday: On your way, you will navigate the hype, lower the risks, and increase your return on investment. Table C.1 summarizes these in a high-level checklist.

It Is Not Over: Exponential Technologies Are Next

We are done, are we not? Well, not really. We never are.

TABLE C.1   Digital Transformation Payday "checklist."

Topic	Considerations
Digital Transformation Payday	• Beware of the digital hype trap and the numerous forces in the digital ecosystem that do not necessarily always have only your payday in mind.
	• There are no such things as easy solutions, universally winning recipes, or directly transferrable digital transformation best practices.
	• If you want your payday to happen with lower risks, never believe in ready-made checklists, maturity models, or business cases. Listen to them, read them, digest them, and then do your own thing based on your real-life context.
	• Ultimately, only you can define and judge all transformation elements from our framework and their corresponding accelerators/decelerators end-to-end.
	• You are also the only one who can make sure that your return on investment will materialize and your payday will happen.
Design/strategy	• You must develop a winning strategy.
	• If you do not, what you end up doing will be your strategy, likely with undesired outcomes.
	• Without a defined strategy to win, it's better to not even start your transformation because it will likely not lead to any sustainable competitive advantage and the related paydays you are hoping and striving for.
Catalysts/drivers	• Remember, digital transformation is not all about new technology supply. Technologies are heavily interdependent and all by themselves do not scale without new capabilities and funding.
	• Do not underestimate your customers' and workforces' rapidly changing expectations; they expect nothing less from you.
	• Always watch out for threats and opportunities from blurring industry boundaries.
Reactants/scope	• Make sure that whatever you do ends up transforming your core, your likely biggest payday lever in scope.
	• Even if you start at the frontier or in the adjacencies of your business, never forget the core and prepare for later reintegration, transfers, or substitution from the get-go.

(continued)

TABLE C.1	*(continued)*
Reaction mechanism/ process	• Never underestimate what it really takes to bring seemingly intuitive agile or hybrid transformation concepts to life at scale in your specific context. • Agile done incorrectly can be a very effective tool to prevent your payday for infinity.
Product/ outcomes	• Identify new ways to track outcomes and progress against all pain and resistance. • Measure, measure, measure your paydays.
Predictors	• You are what you are. Know and carefully manage the baseline for your payday efforts and expectations (market context, financials, and communication). • They will heavily influence your odds of a successful digital transformation payday loop.

While this book had digital transformations in mind, it is also applicable to any (disruptive) technology-driven transformation. Most likely, the world already is progressing or will soon move to the next stage of development. It is believed to quickly evolve "beyond digital," where so-called "exponentials" (intelligent processes, integrated reality, new energy matrix, digital governance, bioprogramming and neurogamification) make digital transformation look "boring" (Rodriguez-Ramos 2018, pp. 1–9) in comparison.

But that is a different story and maybe worth another book.

PART IV

The Science Behind the Book

APPENDIX A

A Lazy Reader's Guide to Key Digital Transformation Definitions from Practice and Science

As explained before there are more definitions of digital transformation out there than anyone can handle. To make your life easier, Tables A.1 and A.2 cluster the literature from management practice and science into our framework (except for the design/strategy cluster) and raise any professional services affiliation of the authors to better understand the marketing character of each publication.

TABLE A.1 Management practice definitions of digital transformation (selection/own summary).

Management Practice Authors (Chronological)	Digital Transformation Payday Elements (Selection of Highlights Across Sources)			Professional Services Affiliation
	Catalysts/Drivers	Reactants/Scope and Reaction Mechanisms/ Processes	Products/Outcomes	
Kane et al. (2018, Abstract)	"Adapting to the digital market environment and taking advantage of digital technologies to improve operations and drive new customer value" (p. 3).	"[. . .] Digitally maturing companies do more than just run experiments" (p. 11).	Case examples	Yes
Davenport and Westerman (2018)	"It is multifaceted and diffuse and doesn't just involve technology" (p. 4).	"It requires foundational investments in skills, projects, infrastructure and, often, in cleaning up IT-systems. It requires mixing people, machines and business processes" (p. 4).	"No digital initiative is undertaken . . . if it doesn't fit the strategy closely and if it's not hardwired to value" (p. 4).	No
Gale and Aarons (2017)	"Seven drivers that will help you escape the old world" (p. 31ff).	"The digital Helix 8. . . as a framework for success" (p. 103ff).	"The expectation chasm" (p. 54, p. 69ff.) "Digital takes different metrics" (p. 55ff).	Yes
Rogers (2016)	"Many of the fundamental rules and assumptions that governed and grew [. . .] businesses no longer hold" (p. 1).	". . . process of holistic transformation" (p. 18).	"Re-imagining your business" (p. 239).	Yes

Management Practice

Digital Transformation Payday Elements (Selection of Highlights Across Sources)

Authors (Chronological)	Catalysts/Drivers	Reactants/Scope and Reaction Mechanisms/Processes	Products/Outcomes	Professional Services Affiliation
Brynjolfsson and McAfee (2014)	"Technology races ahead" (p. 13ff.). "Moore's law and the second half of the chessboard" (p. 39ff.).	"Digitization of just about everything" (p. 57ff.).	"The biggest winners: Stars and Superstars" (p. 147ff.).	No
Raskino and Waller (2015)	"Every product will be digitally remastered" (p. 23ff.). "Catch the triple tipping point" (technology, regulation, culture" (p. 43ff.). "Monitor industry boundary blurring" (p. 63ff.).	"Remodel your enterprise" (p. 83ff.).	"Change the rules of the game" (p. 2).	Yes
Westerman, Bonnet, and McAfee (2014)	"Technology is . . . bigger because recent progress in all things digital is removing constraints and creating exciting new possibilities" (p. 1).	"Creating a compelling customer experience" "Exploiting the power of core operations" "Reinventing business models" "Building leadership capabilities" (p. v).	"Digital Masters outperform their peers" (EBIT and net profit margin) (p. 18).	Yes

TABLE A.2 Scientific research definitions of Digital Transformation (selection/own summary).

Scientific Research	Digital Transformation Payday Elements (Selection of Highlights Across Sources)			Lit. Review Only	Satisfies Our Defined Digital Transformation Criteria (Chapter 2)
Author (Chronological)	Catalysts/Drivers	Reactants/Scope and Reaction Mechanisms/Processes	Products/Outcomes		
Hund, Drechsler, and Reibenspiess (2019) (however focused on digital innovation, not Transformation)	"Emergence of new industrial structures" (p. 10). "The dawn and subsequent rise of numerous digital innovations" (p. 2).	"Changes in organizational positioning" (p. 11). "Changes in labour relation to digital innovation" (p. 11). "Increased cooperation across boundaries" (p. 11). "Integration of platforms" (p. 11).	"New forms of measuring success" (p. 7).	Yes	No
Pramanik, Kirtania, and Pani (2019)	"Implementing new age technologies" (p. 5).	"Re-imagine possibilities including extension, interaction, convergence, modularization and integration of prevalent business with digital technologies" (p. 5).	"Business benefits derived through digital technology" (p. 13.)	No	Yes

Scientific Research	Digital Transformation Payday Elements (Selection of Highlights Across Sources)				Satisfies Our Defined Digital Transformation Criteria (Chapter 2)
Author (Chronological)	Catalysts/ Drivers	Reactants/Scope and Reaction Mechanisms/ Processes	Products/ Outcomes	Lit. Review Only	
Beutel (2018) (focused on a unique concept of "digital orientation")	Not discussed explicitly	"The use of digital technologies in products and services for the customer to digitize the firms internal and inter-firm processes and infrastructure" (p. 2).	"To achieve competitive advantage" (p. 2). "Analysed with profitability (profit) and market value (Tobin's Q)"	No	Yes
Bohnsack et al. (2018)	"Enablers" (including technologies) "Firm capabilities" "Context" (detailed lists given all on p. 9).	"Leads to manifold changes spanning multiple levels (such as individual, organizational, societal)" "Scope, risks, barriers, trajectory, agency" (detailed lists given all on p. 9).	"Change" "Economics" (detailed lists given all on p. 9).	Yes	Yes
Hinings, Gegenhuber, and Greenwood (2018)	"Combined effects of several digital innovations" (p. 1).	"Bringing about novel actors (and actor constellations, structures, practices, values and beliefs" (p. 1).	"Change, threaten, complement or replace existing rules of the game" (p. 1).	No	Yes

(continued)

TABLE A.2 *(continued)*

Scientific Research Author (Chronological)	Digital Transformation Payday Elements (Selection of Highlights Across Sources)			Lit. Review Only	Satisfies Our Defined Digital Transformation Criteria (Chapter 2)
	Catalysts/ Drivers	Reactants/Scope and Reaction Mechanisms/ Processes	Products/ Outcomes		
Vukšić, Ivančić, and Vugec (2018)	"The use of new technologies (e.g. advanced analytics, machine learning, artificial intelligence applications, Internet of Things)" (p. 737).	"Changes key business elements, including strategy, business model, business processes, organizational structures and organizational culture" (p. 737).	"Can lead to business process optimization and overall better organizational performance" "Triggers industry disruption" (p. 737).	No	Yes
Osmundsen, Iden, and Bygstad (2018)	"Can be seen as external or internal triggers" "To keep up with digital shifts" (p. 5).	"Reformed IS organization" "New business models" "Agility" (p. 8ff).	"Direct or indirect effects on firm outcomes and different performance measures" (p. 10).	Yes	Yes

Scientific Research	Digital Transformation Payday Elements (Selection of Highlights Across Sources)			Lit. Review Only	Satisfies Our Defined Digital Transformation Criteria (Chapter 2)
Author (Chronological)	Catalysts/ Drivers	Reactants/Scope and Reaction Mechanisms/ Processes	Products/ Outcomes		
Schallmo et al. (2017) (translated for this research)	"Leverage new technologies" (p. 5). "Capabilities to generate and analyse data" "Enablers" (p. 5).	"Across all value chain steps" (p. 5). "For firms, business models, processes, relationships, products, etc.) (p. 5). "Incremental or radical process" (p. 5.).	"Financial dimension" (p. 14ff).	No	Partially, not exclusively for firms overall
Nambisan et al. (2017) (focused on digital innovation)	"A broad swath of digital tools" (p. 224).	"Creation (and consequent change in) market offerings, business processes, or models" (p. 224).	"A range of innovation outcomes . . . do not need to be digital" (p. 224).	No	Partially, but focused on innovation only
Sebastian et al. (2017)	"New technologies, particularly what we refer to as SMACIT (social, mobile, analytics, cloud and Internet of Things" (p. 197).	"SMACIT inspired value proposition" (p. 198). "Operational backbone" (p. 201). " . . . digital services platform" (p. 201).	Case examples	No	Yes

(continued)

TABLE A.2 (*continued*)

Scientific Research Author (Chronological)	Digital Transformation Payday Elements (Selection of Highlights Across Sources)				Satisfies Our Defined Digital Transformation Criteria (Chapter 2)
	Catalysts/ Drivers	Reactants/Scope and Reaction Mechanisms/ Processes	Products/ Outcomes	Lit. Review Only	
Hess et al. (2016)	". . . market-changing potential of digital technologies" (p. 123). "Financial pressure on current core business" (p. 138).	"Concerned with the changes digital technologies can bring about in a company's business model, which result in changed products or organizational structure or the automation of processes" (p. 124). "Affects many or all segments within a company" (p. 124). "Across company borders" (p. 125).	Case examples	No	Yes
Henriette, Feki, and Boughzala (2015)	"Industries are facing technology shifts" "Market volatility" "Better response to demand" (p. 2).	A business model driven by the impact of digital technology on society "Business models" "Operational processes" "Experience" (p. 10).	Not discussed	Yes	Yes

APPENDIX B

How to Measure Digital Transformation Efforts in Annual Reports with Dictionary-Based Automated Textual Analysis

Due to the widespread lack of concrete KPIs for measuring digital transformation outcomes, several authors (Beutel 2018; Chen and Srinivasan 2019; Hossnofsky and Junge 2019) have already applied more basic automated textual analysis. They mostly focused on numerical counting of occurrences of some sort of digital dictionary terms in financial reports and/or earnings calls as proxies for digital transformation outcomes. The original research underlying this book followed the same idea by using the previously described "replicable references" It went substantially further, not only by integrating the developed digital transformation transversal framework for clustering of dictionary terms (which was discarded in this book for the sake of simplicity), but, more important, by adding natural language processing (NLP) built on Python, a commonly used, easy-to-learn programming language for machine learning libraries (Python 2020): Mostly NLTK and spaCy (NLTK 2020; spaCy 2020;) for textual analysis beyond mere counting of occurrences/frequencies.

Based on these design decisions, a unique digital transformation language dictionary, across all major framework categories (catalysts, reactants, reaction mechanisms, and products), built the foundation of all subsequent analysis. (See Figure B.1.) To limit the data size explosion and the risk of potentially insufficient numbers of observations, the decision was made to stay on element level (catalysts, reactants, reaction mechanisms, and products) and not go down further below (for example, supply and demand or even down further). The initial version of this proprietary dictionary was compiled, and framework-category clustered manually with a purposely very broad scope of terms. It covered around 400 digital (technology) related words and combined applicable literature research findings (Beutel 2018; Briggs et al. 2019; Hossnofsky and Junge 2019) with my practical experience. This dictionary already has, due to the widespread areas of this category, a strong dominance of catalyst terms. In a second step, recent natural language processing (NLP) advancements were leveraged in three different, state-of-the-art wordembedding/vectoring algorithms implemented by one prepackaged Python module called "Magnitude" (Patel et al. 2018). Two FastText algorithms (Bojanowski et al. 2017) and one ELMo algorithm

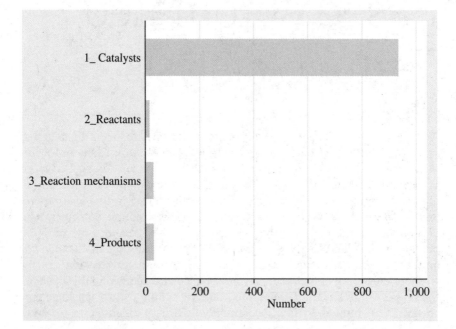

FIGURE B.1 Digital transformation (language) dictionary categories.

(Peters et al. 2018) were selected. This allowed expanding each word in the dictionary with similar terms (15 for each algorithm, leading to a total long-list of 9,346 words after removing duplicates). The third step then aimed to manually clean the resulting longlist from all irrelevant words by a combination of the author's expertise in the field and a supplementary crosscheck of a second digital transformation executive. This joint cleansing process applied several principles: It removed "typo similarities" exposed by the algorithms ("virutal," "softvare") as no such typos are expected in SEC filings. It deleted obvious company names (for example, Apple, Microsoft, Salesforce, etc.) as there are substantial overlaps between chosen portfolio companies as such and the digital transformation related terms as derived by the above algorithms. This prevented distortions in counting occurrences of these companies as a representation of digital transformation in their own and other companies' reports. Next, it removed obvious wrong associations (IT "cloud" versus "thundercloud").

Finally, the chosen approach which possibly eliminated (with custom Python code) all words with the same lemmas ("automate" as the base form versus "automated" as a variation with the same lemma "automate"). This later avoided double counting of lemmatized tokens/words. The final dictionary now covers 1,008 terms in total across all framework categories. The described dominance of catalyst terms is true also for the final version. This does not affect us for this book, as all categories were in the end aggregated to one single measurement only for the sake of simplicity.

APPENDIX C

How to Compile a Unique Financial Database of More than 20,000 Annual Reports

As stated, this book focuses on digital transformation value implications for established capital market listed firms with a sufficient history to require a transformation from a given digital status-quo to a more advanced digital outcome. Consequently, greenfield companies are not in scope. To achieve similar goals, related work has first typically included a broad variety of listed company portfolios, ranging from global surveys (Westerman, Bonnet, and McAfee 2014), to various US companies (Beutel 2018; Chen and Srinivasan 2019) and exclusively Europe-centric approaches (Hossnofsky and Junge 2019; Kawohl and Hüpel 2018; Wroblewski 2018). Second, the authors applied divergent data sources (COMPUSTAT, CRSP, IBES, Thomson Reuters DataStream and annual reports) for financial analysis. Therefore, clear choices were required for these two key research foundations.

Different global stock indexes have carefully been considered to construct the final listed large corporate portfolio for analysis. In the end, after a cautious assessment—the NASDAQ and the New York Stock Exchange (NYSE)—two US-based regional stock exchanges for large corporates turned out to be the best-fitting choice. They were found to provide a good match across all data requirements. First, due to the focus on US-listed companies, geographical market context differences are of less relevance for the planned valuation approach. Second, the chosen companies, due to their listing in the United States, provide a sufficient data history, guaranteed quality, and consistency based on the conservatism in applicable US-GAAP accounting rules. This further supports a good fit to the assumptions required for the later applied clean-surplus-based valuation models (Falkum 2011). Finally, they provide a sufficiently broad industry and sector coverage for better generalizability of all findings. Obviously, depending on the availability of datapoints for each analysis, regressions will only use a limited part of these observations (see Table C.1). For example, because of leveraging the lagged ROA (L1ROA) for our models, the number of observations naturally must go down closer to the 20K range.

The portfolio has yearly on average shown market capitalization (avg-MARKETCAP) growth, the expected COVID19 dip in 2020 and fluctuating negative mid-term ROA performance (ROA3Y) over the full period as can be seen in Figures C.1 and C.2.

I spare you the required endless tables here, but rest assured that the final dataset is adequately balanced across sectors and industries, covering eight sectors, 31 industry categories, and 206 industry groups. This is true when reviewing all observations from the dataset as well as when assessing only the relevant data subset for financial analysis, that is, after removing all observations eliminated due to missing financial data in the final analysis.

From day one of starting the research, the vision was to develop a harmonized and single "source of truth" including a solution for the difficult timing congruence of textual and financial information and the ability to cor-

TABLE C.1 **Overall sample financial summary statistics.**

Financial variables	N	Mean	SD	Min	Max
MARKETCAP	27522	1.64E+10	1.11E+12	-2.38E+11	1.82E+14
ROA3Y	20997	-0.0582299	0.927666	-53.67982	13.17412
TOTAL EQUITY	29886	2.76E+09	1.22E+10	-1.83E+10	3.10E+11
NET INCOME	30064	3.47E+08	1.95E+09	-2.72E+10	5.95E+10
AOCI	30544	-1.60E+08	1.26E+09	-5.00E+10	1.81E+10
PAYMENT OF DIVIDENDS	30544	-1.67E+08	8.49E+08	-6.67E+10	4.54E+09
DELTA EQUITY	30544	2.80E+07	5.00E+08	0	3.07E+10
L1ROA	24942	-1.076136	159.0729	-25120	190.8323
NET DEBT	29931	1.19E+09	1.53E+10	-4.97E+11	5.61E+11
INVESTED CAPITAL GROWTH	28152	130.0668	10427.98	-15714.08	1182771
BOOK TO MARKET	26948	6.188369	534.8125	-4918.638	82220.16

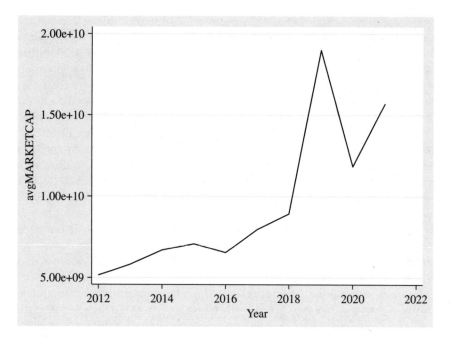

FIGURE C.1   Portfolio average market capitalization 2012–2021 (in USD).

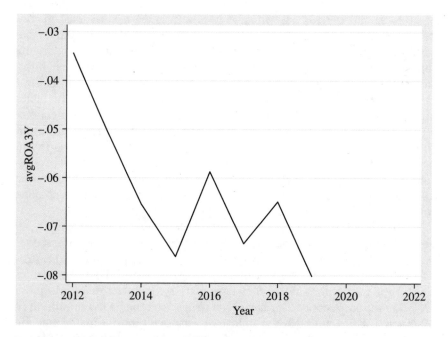

FIGURE C.2   Portfolio average ROA3Y 2011–2021 (in percent).

rectly match these two elements. The aim was to create a unique data source never before available for digital transformation research. After looking into different automated and manual data sourcing options applied in literature (Beutel 2018; Chen and Srinivasan 2019; Kawohl and Hüpel 2018; Hossnof-sky and Junge 2019; Cohen, Malloy, and Nguyen 2020), it became clear that getting data as close as possible from its source would be most efficient for the planned advanced analysis. Fortunately, the US Securities and Exchange Commission's (SEC's) HTTPS file system allows scraping the "Electronic Data Gathering, Analysis, and Retrieval system" (EDGAR) filings by corpora-tions, funds, and individuals. "EDGAR indexes list all public SEC filings for each quarter starting in the third quarter of 1994 to the present" (SEC 2020a). Scraping data directly from EDGAR provided three clear benefits. First and foremost, all relevant companies were addressable, with reduced need to exclude companies for data availability reasons. Second, EDGAR provided electronic accessibility via flexible APIs including the option to export full text and not just PDFs in a unified structure and content classification across all reports. Third, matching market, value and other complementary financial data could be sourced from one specialized financial information data source: Intrineo. Intrineo data feed also directly builds on EDGAR, reaches back until 2007 and has relevant data for this research purpose from 2009 (Intrinio 2020) as further laid out later.

Out of the substantial range of reports available per company (SEC 2020b), only the overall "bulk" content of standardized annual filings (10-K, 10-K405 and the 10-KSB, which is irrelevant here) was assumed to con-tain relevant information. This is fully in line with conceptually comparable textual analysis work (Cohen, Malloy, and Nguyen 2020) plus its underlying foundational work (Loughran and McDonald 2011). As in this research, infor-mation was further processed in a first stage by removing "clutter" (numerical tables with > 15% of numerical characters, HTML tags, newlines, XBRL tables and Unicode text) to produce a "raw" form of all reports. Underlying chapter structures of the processed reports were therefore deemed irrelevant for all further analysis. As mentioned earlier, the applied portfolio list was frozen at a fixed cut-off date, end of June 2021, to compile the final portfolio. A cus-tom developed Python (Python 2020) code then scraped all corresponding 10-K/10-K405 filings for each filing date, going back in time as far as possible for the respective companies or 2011, whichever was later.

SEC has published clear rules for submission (SEC 2020c). Originally, all companies had to submit their 10-Ks within 90 days after the fiscal year end-ed. This was changed in 2004, when the SEC approved a new rule that adjust-ed this target to 60 days for so-called "accelerated filers." These companies must fulfill four criteria. First, they have market capitalization of minimum USD 75 million. Second, they have been obliged to report for a minimum of 12 months. Third, they have to upload at least one report. Finally, they are not allowed to submit their reports on Forms 10-QSB and 10-KSB. A further category of "large accelerated filers," with a public float of more than USD 700 million with a deadline of 60 days and a revised final date for "accelerated filers" of 75 days was created later. 10-Q reports, however, are due 45 days after the quarter-end for "nonaccelerated" filers and 40 days after the quarter-end for "accelerated" and "large accelerated" filers (SEC 2020c). This information is relevant for the following valuation analysis insofar as the understanding of potentially significant timing lags after closing and until filing have been implemented as a control, depending on actual information leakage in the meantime (earnings calls). All firms assessed conceptually fall under the "accelerated filer" and "large accelerated filer" categories, which helps pragmatically matching textual and financial data due to a limited time window for reporting.

The specialized financial information platform Intrineo serves as the predominant data source for market, value, and other complementary financial data in this research. This source differs from the more classical data originations typically applied in scientific research (mostly COMPUSTAT, CRSP, IBES, Thomson Reuters DataStream and original annual reports). Intrinio data feed was explicitly chosen because it is largely based on the same SEC EDGAR filings and therefore is consistent with the directly scraped textual EDGAR data (Intrinio 2020). It can serve as the backbone for all market and financial information over a period of 10 years. The combination of direct EDGAR scraping and the Intrinio platform therefore allowed devising the unique empirical approach at the core of this research project. Intrinio data is both accessible as bulk download and via advanced API. The major advantage of the chosen data sourcing approach is the limited need of adding additional data sources. The only exception is yahoo!finance (yahoo 2020), which is only and exclusively used to capture earnings announcements dates to later control for the time lag between these announcements and the official filing of financial information to the SEC database, but it needs to be interpreted with great care due to suspected issues in the data quality delivered with a number of outliers.

Overall, the described customized portfolio of established listed large corporations and the combination of consistent data sources—namely, SEC's EDGAR database for textual data and the Intrinio platform for most financial data—have proven to be the best fit for the major research requirement: a reliable and cross-industry representative data set for analyzing the value impact of digital transformation on larger corporations.

APPENDIX D

How to Start Understanding What the Annual Reports Say About Digital with the Help of Natural Language Processing (NLP)

To leverage the planned NLP concepts into the empirical approach, I developed a new concept of what I call "quantification levels." The idea behind this concept is to assign higher "levels" the more specific/concrete the outcome is assumed to be. For this book, I ended up using only both extreme ends (Level 1 and Level 3), but in the original research, a careful second-stage preprocessing of all reports built the foundation for the analysis along all three levels. (See Figure D.1.)

Level 1 analyzed separately for each 10-K filing date the amount of occurrences/frequency of digital transformation language dictionary terms per category in the "raw" reports and, for normalization purposes, the relative percentage of occurrences versus total words as a proxy for digital transformation outcomes. In simple words, it counts how often digital terms appear in a report and makes this number comparable across reports by adjusting it for document length.

Level 2 (not used for this book) analyzed separately for each 10-K filing date the number of occurrences of digital transformation language dictionary

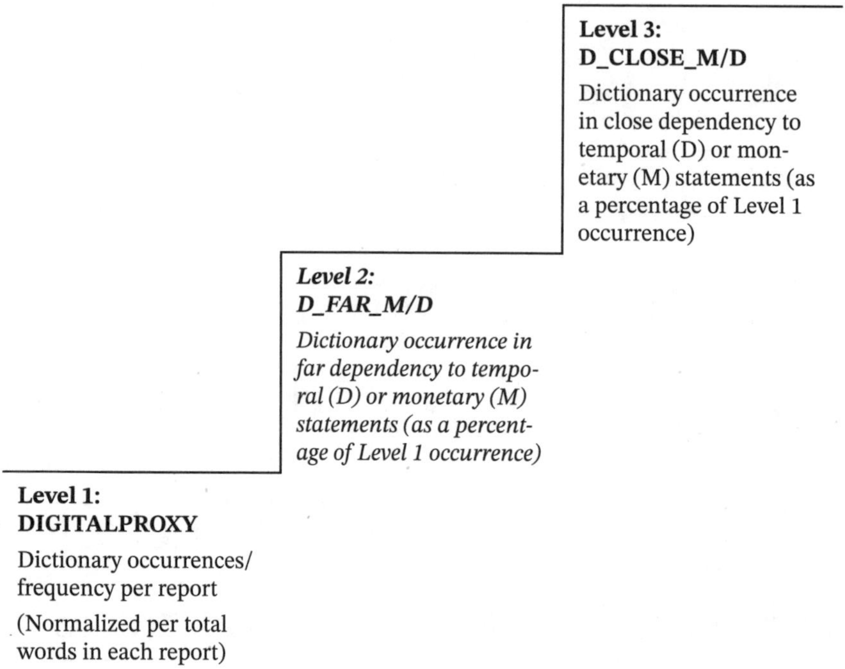

Level 3:
D_CLOSE_M/D

Dictionary occurrence
in close dependency to
temporal (D) or mon-
etary (M) statements (as
a percentage of Level 1
occurrence)

Level 2:
D_FAR_M/D

*Dictionary occurrence in
far dependency to tempo-
ral (D) or monetary (M)
statements (as a percent-
age of Level 1 occurrence)*

Level 1:
DIGITALPROXY

Dictionary occurrences/
frequency per report

(Normalized per total
words in each report)

FIGURE D.1   Replicable references on three analysis levels.

terms per category in relationship to explicit statements on monetary or timing impacts and the relative percentage of these occurrences (versus total
level 1 occurrences) as a proxy for digital transformation outcomes. In simple
words it counts the percentage of Level 1 terms that are further specified by
dates or monetary terms in far proximity to the terms (measured in so-called
arcs, that is, dependency steps in the sentence syntax until you find the specification). In Level 2 only syntax dependencies above four arcs in the same
sentence are counted.

Level 3 analyzed separately for each 10-K filing date the number of occurrences of digital transformation language dictionary terms per category in relationship to explicit statements on monetary or timing impacts and the relative
percentage of these occurrences (versus total Level 1 occurrences) as a proxy
for digital transformation outcomes. In simple words it counts the percentage
of Level 1 terms, which are further specified by dates or monetary terms in
close proximity to the terms (measured in so-called arcs, that is, dependency
steps in the sentence syntax until you find the specification). In Level 3 only
close syntax dependencies below five arcs in the same sentence are counted.

For the sake of simplicity in this book, the sublevels of framework categories, which would be available in the dataset have been discarded, and only
aggregate figures across all framework categories are applied.

After random checks, it became clear that programmatically eliminating obvious company names would be beneficial. This was implemented with an entity search algorithm.

In the wake of the preceding application of NLP methods via custom-automated Python code, it turned out to be very efficient to leverage further supplementary analysis to generate additional information on the assessed texts. This builds on a recent trend in literature to look deeper into text sentiments to derive further conclusions and/or controls for subjectivity/objectivity and negative/positive statements in SEC 10-K reports versus their financial impact, usually with a confirmation of a relationship between financials and sentiment data (Chouliaras 2015; Li 2006). In order to generate a first operationalization of the desired text sentiments, a custom-developed Python code was applied, strongly building on similar lexical development work (Haritash 2018). The general idea of this pragmatic approach is simply counting words from a "negative" dictionary, as the negative score (S_N) (for example "annulments," "annuls," "anomalies," "anomalous"), then counting words from a "positive" dictionary, as the positive score (S_P) (for example, "able," "abundance, "acclaimed," "accomplish") (UND 2020) and then calculating a so called "polarity" (S_{POL}) out of the relation of these two scores. This score determines if a given text is positive or negative in nature. It is calculated by using the following formula (range is from –1 to +1):

$$S_{POL} = \frac{S_P - S_N}{S_P + S_N + 0.000001}$$

In addition to the custom-developed approach the underlying research for this book also used "TextBlob," which ended up as the only tool for all sentiment analysis in this book. TextBlob is a Python library for processing textual data. "It provides API for natural language processing (NLP) tasks such as part-of-speech tagging, noun phrase extraction, sentiment analysis, classification, translation . . ." (Loria 2018, website). This work, however, only

TABLE D.1 Overall sample textual analysis summary statistics.

Textual Variables	N	Mean	SD	Min	Max
DIGITALPROXY	30544	2.480327	2.819965	0	30.91132
D_CLOSE_M	30540	0.002627	0.008951	0	0.193548
D_CLOSE_D	30540	0.002943	0.009139	0	0.166667
POLARITY	30544	0.051735	0.013579	-0.0197173	0.141112
SUBJECTIVITY	30544	0.374910	0.018039	0.2806215	0.459227

SD = Standard deviation

leverages the prepacked sentiment functionality to calculate subjectivity and polarity scores for the generated "raw" texts.

In summary, applied textual analysis produced several variables (see Table D.1) sourced via custom-developed Python code directly from the pre-processed 10-Ks.

Obviously, depending on the availability of datapoints for each analysis, regressions will only include a limited part of these observations, for example, because of leveraging the lagged ROA (L1ROA) for our models, the number of observations naturally must go down closer to the 20,000 range.

APPENDIX E
How to Link Digital Transformation and Value with the Residual Income Valuation Model

The main philosophy behind all conducted financial analysis was to apply a balanced mix of scientific rigor and practitioner-minded pragmatism for operationalization. This implied no distractions by specialized valuation research models and helped to keep track of accessible digital transformation data, instead of dwelling into theoretical peculiarities of detailed model parametrization. Financial valuation methods have been grouped in many different ways, depending on their objective and approach (Falkum 2011; Vartanian 2003). As indicated earlier, direct and foundational digital transformation, technology/IT/IS, innovation, and corporate finance-value research over time have applied a seemingly "infinite" variation of customized approaches out of this basket of potential valuation analysis.

From a practitioner perspective, further reviewing this substantial breadth of models was of no further interest for this research, as it has already been extensively analyzed in corresponding meta-studies, for example, for IT-payoff (Kohli and Devaraj 2003; Sabherwal and Jeyaraj 2015) or for innovation-payoff (Vartanian 2003) and, after careful consideration, added no further relevant insights. Instead, the much more promising angle was the subset of approaches, which found theoretically sound ways to integrate non-accounting "other information" into models otherwise fully based on publicly available financial reporting. After a careful assessment, residual income valuation models, and specifically the Ohlson model (Ohlson 1995, 2001), seemed to be the best starting point. Since being introduced by Ohlson in the 1990s, residual income valuation (RIM) models have opened new possibilities to leverage accounting-based information in valuation. They have been shown to produce better results than the previously favored cash-based models (Gao et al. 2019; Ohlson 2001). Nevertheless, in their basic form, they were not without criticism, often with the reproach of systematic undervaluation, as will be elaborated later in more detail.

So how does it work? As described by McCrae and Nilsson (2001), whose explanations are used as the basis for summarizing the majority of theoretical foundations in the following, two major assumptions are required for any basic Ohlson-related model to be valid. First, the market value of a firm's equity (P_t) equals the present value of future dividends payments $(d_{t+\tau})$. Equity contributions are included as negative dividends. This implies:

$$P_t = \sum_{\tau=1}^{\infty} (1+r)^{-\tau} E_t \left[d_{t+\tau} \right]$$

In this case (r) equals the cost of equity capital and $(E_t[.])$ is used as the expectation operator based on available information at time (t).

The second assumption calls for the book value changes over time to not oppose the so-called clean surplus principle. This implies that any change in book value, period to period, is equal to the earnings minus net dividends, resulting in:

$$BV_t = BV_{t-1} + x_t - d_t$$

Here (BV_t) equals book value at time (t), (x_t) the term for earnings for period (t), and (d_t) shows the net dividends given to shareholders at time (t). All valuation-relevant information needs to be at some point reflected in the profit and loss statements to fulfill the requirements of lean surplus accounting. This establishes the so-called residual income or abnormal earnings (x_t^a):

$$x_t^a / x_t - rBV_{t-1}$$

These two equations combine in the valuation function:

$$P_t = \sum_{\tau=1}^{\infty} (1+r)^{-\tau} E_t \left[x_{t+\tau}^a \right]$$

Ohlson's major theoretical contribution was an extension to this equation: His "linear information dynamics" (LIM) solved the limitation that it is only based on future values and thus does not allow to link publicly available accounting figures to equity value. For Ohlson, the information of expected future abnormal earnings was based on both the history of abnormal earnings and non-accounting "other information" not reflected in this history (ϑ). He then postulated a modified first-order autoregressive process for these abnormal earnings and a simple first-order autoregressive process for non-accounting information.

This gives for abnormal earnings (or residual income):

$$x_{t+1}^a = \omega x_t^a + \vartheta_t + \varepsilon_{1,t+1}$$

For non-accounting "other information," this implies:

$$\vartheta_{t+1} = \gamma \vartheta_t + \varepsilon_{2,t+1}$$

Here (ω) and (γ) are fixed persistence parameters that are supposed to be transparent, greater than zero, and less than one. The variable (ϑ_t) represents non-accounting information on expected future abnormal earnings that is seen at the end of period (t) but not yet reflected in accounting, and (ε) represents a random error term assumed to be mean zero and uncorrelated over time.

Ohlson's other important contribution was to fix the finite valuation problem embedded in most traditional models: Assuming a (ω) and (γ) of below one, abnormal earnings converge to zero, so that in the long run the firm's book and market value will converge. This fits to the theoretical notion that efficient market competition should eventually cancel out firm-specific abnormal earnings. This leads to a valuation function for market value based on abnormal earnings, accounting book values and non-accounting information:

$$P_t = BV_t + \alpha_1 x_1^a + \alpha_2 \vartheta_t$$

where:

$$a_1 = \omega / (1 + r - \omega)$$

$$a_2 = (1+r) / (1 + r - \omega)(1 + r - \gamma)$$

This new theoretical approach finally laid a foundation for applying accounting data to explain and predict market values, where shareholder value consequently has three major elements: the current book value, the capitalized current residual income, and the capitalized value implied by non-accounting "other information." Investors are, in other words, presumed to trade current asset value for a future stream of expected income. Correspondingly, asset prices embody the present value of all future dividends expected. The Ohlson model ". . . replaces the expected value of future dividends with the book value of equity and current earnings . . . based on the already explained clean surplus principle, which holds that the change in book value of equity will be equal to earnings less paid out dividends and other changes in capital contributions" (Muhanna and Stoel 2010, p. 50).

In operational implementation terms, this leads to the following regression:

$$
\begin{aligned}
\text{MARKETCAP}_{jt} \\
= b_0 + b_1\text{DIGITALPROXY}_{jt} + b_2\text{TOTALEQUITY}_{jt} + b_3\text{NETINCOME}_{jt} \\
+ b_4\text{ACOI}_{jt} + b_5\text{PAYMENTOFDIVIDENDS}_{jt} + b_6\text{DELTAEQUITY}_{jt} \\
+ b_7\text{REVENUEGROWTH}_{jt} + b_8\text{ROA}_{jt-1} \\
+ b_9\text{DELTAEARNINGSDATE}_{jt} + b_{10}\text{POLARITY}_{jt} \\
+ b_{11}\text{SUBJECTIVITY}_{jt} \\
+ b_{12}\text{BOOKTOMARKET}_{jt} + b_{13}\text{INVESTEDCAPITALGROWTH}_{jt} \\
+ \text{firm}\,|\,\text{industry} + \text{interactions} + \text{year} + e_{jt}
\end{aligned}
$$

$$=$$

where (MARKETCAP_{jt}) is the fiscal year-end equity market value for firm (j) for year (t), $(\text{DIGITALPROXY}_{jt})$ is our proxy variable, $(\text{TOTALEQUITY}_{jt})$ is the fiscal year-end book value of equity, (NETINCOME_{jt}) is net income, ACOI_{jt} is accumulated other comprehensive income, and $(\text{PAYMENTOFDIVIDENDS}_{jt})$ and $(\text{DELTAEQUITY}_{jt})$ are dividends paid minus changes in contributed capital, derived from sales of common and preferred stock minus purchases of common and preferred stock. All other parameters are self-explanatory financial key performance indicators (KPIs) or sentiment parameters as controls except for $(\text{DELTAEARNINGSDATE}_{jt})$, which describes the difference in days between the earnings call and the following submission of information to the SEC. Differences in market values beyond book value and earnings are reflected in (BOOKTOMARKET) and (e_{jt}) is the error term. Year and firm|industry indicate the found time and firm/industry fixed effects, interactions the possible variable interactions.

The clean surplus rule, as just elaborated, is one key assumption underlying the applicability of the Ohlson model. Even though US-GAAP, the applicable accounting standards used for all financial information sources in this research, are seen as superior to European accounting standards in this regard, they still include several violations (Falkum 2011; Lo and Lys 2000). Fortunately, FASB Statement No. 130, "Reporting Comprehensive Income," partially mitigates this, as it requires transparency of all income-neutral but equity-influencing actions in the financial statement, so that overall the assumption of clean surplus validity seemed to be fair to make (Falkum 2011) also for the sake of this research, when including this information.

Due to its wide application and acceptance, residual income models (RIM) like the chosen Ohlson model (OM) enjoy the benefit of having been

TABLE E.1 Residual income model research overview.

Author	Research Focus	Results
Barth et al. (1999)	Implementation and empirical tests of different variants	Better results with decomposed income models
	Analysis of industry impact	Industry-specific parameters provide better results
Biddle, Chen, and Zhang (2001)	Expansion of OM	Investments correlate with investment return
	Empirical tests of different hypothesis	Convexity of current and future residual incomes
Choi, O'Hanlon, and Pope (2003)	Expansion of OM by newly developed conservatism	Approach improves OM model
Dechow, Hutton, and Sloan (1999)	Implementation of Ohlson model (OM) supplemented by analyst forecasts	Tendency of undervaluation
	Using different persistence parameters	Analyst forecasts have strong influence on investors expectation building
McCrae and Nilsson (2001)	Country specifics	Significant differences in results between US and non-US firms (Sweden)
Myers (2000)	Implementation of OM without other information and variations	All methods tend to undervalue
		Tendency to project too low residual earnings
Ota (2002)	Empirical tests of different variants	Tendency of undervaluation
		Analyst forecasts help to explain market values and investor expectation building

very broadly tested empirically over the past years (Callen and Segal 2005) as summarized in Table E.1. While "accounting based valuation models studies of US firms tend to support Ohlson's proposition that residual and book value numbers have information content in explaining observed market values" (McCrae and Nilsson 2001, p. 315), potential weaknesses are also very transparent. This enables a much more educated discussion of empirical results compared to other, more custom models.

Implications for this book are manifold. Primarily, the repeating theme of undervaluation leads to the necessity of carefully interpreting results. Furthermore, the value of adding analyst forecasts seems to be visible. Nevertheless, the author has, for the sake of easier operationalization, decided to stick to a simpler version of the Ohlson model, noting the inclusion of

TABLE E.2 Model operationalization overview.

Author	Underlying Research Field	Model Option Description	Evaluation
Anderson, Banker, and Ravindran (2006)	IT value	Custom model	Theoretical foundation available Less relevant as focused only on Y2K event
Beutel (2018)	Digital transformation value	Custom model	Theoretical foundation available All required variables available in chosen data sources but different focus
Chen and Srinivasan (2019)	Digital transformation value	Custom model	Theoretical foundation available Medium feasibility of operationalization due to high complexity
Hossnofsky and Junge (2019)	Digital transformation value	Custom model	Theoretical foundation available High feasibility of operationalization but different focus (analyst view)
Mani, Nandkumar, and Bharadwaj (2018)	Digital transformation value	Custom model	Theoretical foundation available High feasibility of operationalization but different focus
Mithas, Ramasubbu, and Sambamurthy (2011)	IT value	Custom model	Theoretical foundation available Only questionnaire based
Muhanna and Stoel (2010)	IT value	Basic Ohlson model	Strong theoretical foundation All required variables available in chosen data sources
Vartanian (2003)	Innovation value	Custom model	Strong theoretical foundation Innovation-centric variables

analyst expectations as a further potential improvement for later research. Industry relevance has been taken care of insofar as all "other information" variables applied in this research (digital transformation textual analysis-based variables) are mostly industry-, or even company-specific, including the application of firm and industry fixed effects.

Muhanna and Stoehl's basic Ohlson model operationalization (Muhanna and Stoel 2010) served as a foundation for this research after intense consideration mainly for four reasons. First, it is based on a valid theoretical background as demonstrated. Second, its strong requirement for the clean surplus rule, which otherwise is not easy to meet for all international accounting standards, here can as already discussed be assumed to match with "United States Generally Accepted Accounting Principles" applicable (US GAAP) for the selected companies. Third, the transparency on its weaknesses, based on previous authors' extensive empirical testing, allowed having a better sense for the interpretation of results. Fourth, a final, careful evaluation of potential alternative models, from foundational IT value and innovation value research, identified no better match in terms of theoretical foundation and feasible, thesis-objective-supporting operationalization of variables.

While the market value impact of digital transformation is by design of major interest for this research, it seemed advisable to look at least into one more directly accounting-driven parameter as well. Future earnings were the most obvious choice, given their relevance also in the market value considerations. Instead of developing our own customized model for future earnings analysis, it seemed best to again follow Muhanna and Stoel (2010) and their approach. Other than for the Ohlson model, no extensive empirical tests for the chosen mixed fundamental model exist. Therefore, potential shortcomings have to be fully addressed with our own statistical robustness tests when discussing empirical results. Because no logical or empirical results dictated otherwise, the same approach as selected for residual income regressions was also applied for the mixed fundamental model. To later test the hypothesis on the link of digital transformation with lagged (future) earnings, the herewith applied earnings prediction model fully followed the ideas of Muhanna and Stoel (2010). It combines elements of other earnings prediction approaches used in the literature. The custom final model for future earnings accounted for firm assets plus the historical earnings for the assets. It used the average of return on assets ($ROA3Y$) over three years as the measure of future earnings to allow for a possible lag between digital transformation parameters and the realization of potential value.

$ROA3Y$

$$= b_0 + b_1 \text{DIGITALPROXY}_{jt} + b_2 \text{TOTALASSETS}_{jt} + b_3 \text{NETINCOME}_{jt}$$
$$+ b_4 \text{NETINCOMEGROWTH}_{jt} + b_5 \text{REVENUEGROWTH}_{jt} + b_6 \text{POLARITY}_{jt}$$
$$+ b_7 \text{SUBJECTIVITY}_{jt} + b_8 \text{DELTAEARNINGSDATE}_{jt} + \text{firm} | \text{industry}$$
$$+ \text{interactions} + \text{year} + e_{jt}$$

where all variables are self-explanatory for firm (j) in year (t), and year and firm|industry indicate the found time and firm/industry fixed effects, interactions the possible variable interactions, and (e_{jt}) is the error term.

In summary, financial analysis produced several variables, sourced mostly via Intrinio APIs with by design self-explanatory abbreviations in line with the equations explained earlier. To ensure normality of the dependent variables, both applied variables (MARKETCAP and ROA3Y) were checked for normality, and afterwards natural logged in the final models to account for the identified non-normality. Other transformation methods like the two-step or winsorizing approaches as proposed in related literature (Templeton and Burney 2017) were disregarded due to the fact that already simple natural log normalization provided satisfactory results.

APPENDIX F
Supplementary Analytics for the Fearless

T his appendix serves as the backup for the analysis in the main section of the book and adds additional details, which were deemed too complex for the main flow of the book.

The Orthoforest Causal Estimate Approach

Fortunately, based on the advancements of econometric models, we can build on an orthoforest-based causal estimate model to run detailed elasticity analysis for DIGITALPROXY, logMARKETCAP and more.

What we are doing in simple words is this:

1. Build a generic causal map (constructed from the dataset by leveraging the Python DoWhy library (Sharma and Kiciman 2020)) based on our chosen Ohlson model variables for logMARKETCAP (see Chapter 3) or their extensions (see the different angles of analysis throughout Part III) including dummy variables to reflect fixed industry and time effects.

2. Generate the corresponding causal estimand (also known as our coefficient in the other models) by once more applying the Python DoWhy library (Sharma and Kiciman 2020).

3. Leverage DoWy's interface to the EconML Python library (Oprescu et al. 2019) to run a fully nonparametric orthoforest for this estimand and calculate a mean estimate.

4. Run so-called refutation tests to validate the robustness of the model (Sharma and Kiciman 2020), which are described further in the next section of this appendix.

5. Add elasticity analysis for each model variable in relation to DIGITAL-PROXY on our dependent variable logMARKETCAP by estimating elasticities within the variables' data range.

6. Plot these elasticities to identify potential patterns.

Indications for Causality

To get a better feeling about whether our main analysis is merely indicating correlations or also demonstrating some initial signs of causality, out of the vast range of possible analysis, three robustness tests were chosen.

- Sensitivity analysis
- Instrumental variable analysis
- Empiric dynamic modeling (EDM) to check for effect direction

For *sensitivity analysis* of our basic regressions, a software package called sensemakr (Cinelli, Ferwerda, and Hazlett 2020) available as a supplement to STATA was chosen and manually customized to replace the original simple regressions in the package by our fixed time/industry effect model:

> sensemakr can compute sensitivity statistics for routine reporting, such as the robustness value, which describes the minimum strength that unobserved confounders need to have to overturn a research conclusion. The package also provides plotting tools that visually demonstrate the sensitivity of point estimates and t-values to hypothetical confounders. Finally, sensemakr implements formal bounds on sensitivity parameters by means of comparison with the explanatory power of observed variables. All these tools are based on the familiar "omitted variable bias" framework, do not require assumptions regarding the functional form of the treatment assignment mechanism nor the distribution of the unobserved confounders, and naturally handle multiple, non-linear confounders. With sensemakr, users can transparently report the sensitivity of their causal inferences to unobserved confounding, thereby enabling a more precise, quantitative debate as to what can be concluded from imperfect observational studies. (Cinelli, Ferwerda, and Hazlett 2020, Abstract)

As to be expected with our noisy real-life observational data, results are somewhat mixed. While the contour plot for the coefficient remains positive (Figure F.1) even when adding an unobserved variable up to three times the effect of the chosen benchmark (TOTALEQUITY), it almost

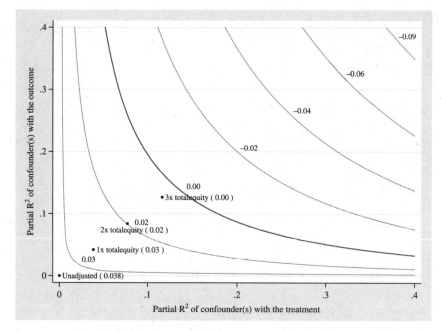

FIGURE F.1   Sensitivity of adding an unobserved variable on coefficient.

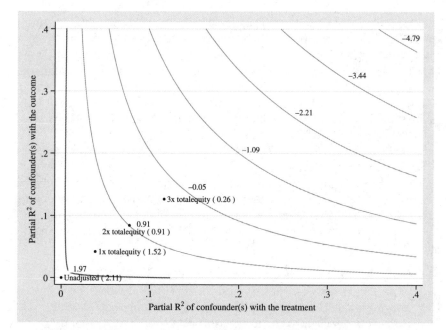

FIGURE F.2   Sensitivity of adding an unobserved variable on significance (t).

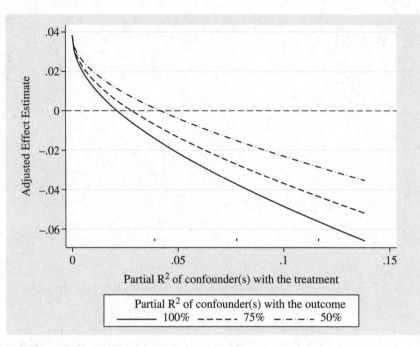

FIGURE F.3 Simulation of extremes.

immediately loses its statistical significance in the corresponding t-contour plot (Figure F.2) with the same benchmark. This can also be seen in the third plot, showing the extremes (Figure F.3).

For the most important sensitivity analysis in our orthoforest estimates, we used the embedded refutation functionalities from DoWhy/EconML (Oprescu et al. 2019; Sharma and Kiciman 2020). Here results are more promising (Table F.1), as all tests show the required behavior, but still obviously cannot guarantee a certain causal effect.

While the search for perfect instrumental variables as a robustness test is something many researchers love, in real life a proper variable to use often does not exist. To follow the pragmatic approach of this book, instrumental variables were only checked generically (Table F.2), using a corresponding STATA package, demonstrating robustness of our model in terms of coefficient and significance even when adding an automatically generated instrumental variable:

"ivreg2h estimates an instrumental variables regression model providing the option to generate instruments using Lewbel's method. This technique allows the identification of structural parameters in regression models with endogenous or mismeasured regressors in the absence of traditional identifying information, such as external instruments or repeated measurements. Identification is achieved in this context by having regressors that are uncorrelated with

the product of heteroskedastic errors, which is a feature of many models where error correlations are due to an unobserved common factor. Using this form of Lewbel's method, instruments may be constructed as simple functions of the model's data. This approach may thus be applied when no external instruments are available, or, alternatively, used to supplement external instruments to improve the efficiency of the IV estimator. Supplementing external instruments can also allow Sargan-Hansen tests of the orthogonality conditions or overidentifying restrictions to be performed, which would not be available in the case of exact identification by external instruments."
(Baum and Schaffer 2021, Abstract)

TABLE F.1   **Refutation sensitivity results.**

Refutation Test	Required Behavior	Estimate	New Estimate	Interpretation
Add a random common cause	The new estimate does not change drastically with a random common cause p-value <0.05	0.0861773	0.0884010 $p = 0.15$	Stable: Estimate seems to be robust to adding another common cause variable p-value >0.05 indicates that based on the test there is no problem with the estimator
Add an unob-served common cause	The new estimate does not change drastically with a small, simulated effect of unob-served common causes	0.0861773	0.0889128 p not applicable	Stable: Estimate seems to be robust to adding another common cause variable
Use a placebo treatment	After the data change, the causal true esti-mate should be zero. Any estimator whose result var-ies significantly from zero on the new data fails the test p-value <0.05	0.0861773	−0.0003231 $p = 0.47475$	Estimate trends to zero as required when a placebo treatment is used p-value >0.05 indicates that based on the test there is no problem with the estimator

(continued)

TABLE F.1	(*continued*)			
Use a sub-set of data	The new estimate does not change drastically with a data subset *p*-value <0.05	0.0861773	0.0860033 *p* = 0.314812	Stable: Estimate seems to be robust to adding another common cause variable, *p*-value >0.05 indicates that based on the test there is no problem with the estimator

TABLE F.2 Instrumental variable robustness check for causality indications.

Models	(7) Industry/Time Fixed Effects Generated Instrumental Variable
Variables	**logMARKETCAP**
DIGITALPROXY	5.136e-02**
	(2.075e-02)
TOTALEQUITY	1.896e-11***
	(2.213e-12)
NETINCOME	3.776e-11***
	(5.821e-12)
AOCI	-1.426e-11***
	(5.208e-12)
PAYMENTOFDIVIDENDS	-1.943e-11***
	(6.247e-12)
DELTAEQUITY	1.141e-04**
	(5.12e-05)
REVENUEGROWTH	4.88e-05***
	(1.25e-06)

TABLE F.2 (*continued*)	
L1.ROA	1.750e-12**
	(7.991e-13)
NETDEBT	7.107e+00***
	(1.056e+00)
INVESTEDCAPITALGROWTH	-4.704e+00***
	(9.416e-01)
BOOKTOMARKET	-1.011e-04***
	(2.59e-05)
POLARITY	5.136e-02**
	(2.075e-02)
SUBJECTIVITY	1.896e-11***
	(2.213e-12)
EARNINGSDATEDELTA	3.776e-11***
	(5.821e-12)
Centered R-squared	0.2123
Uncentered R-squared	0.2123
Basic Fixed Effect	INDUSTRY
Time Fixed Effect	YEAR
Clustered Standard Error	INDUSTRY
Observations	22247

Instrumented: DIGITALPROXY

Included instruments: TOTALEQUITY, NETINCOME, PAYMENTOFDIVIDENDS, DELTAEQUITY, REVENUEGROWTH, L1ROA, NETDEBT, POLARITY, SUBJECTIVITY, DELTAEARNINGS, date, years 2011–2019

For our last robustness test, we leverage *empiric dynamic modeling (EDM)*, a very recent experimental innovation in causal analysis to check for indications of effect direction. This has also been implemented as a package in STATA:

"How can social and health researchers study complex dynamic systems that function in nonlinear and even chaotic ways? Common methods, such as experiments and equation-based models, may be ill-suited to this task. To address the limitations of existing methods and offer nonparametric tools for characterizing and testing causality in nonlinear dynamic systems, we introduce the edm command in STATA. This command implements three key empirical dynamic modeling (EDM) methods for time series and panel data: 1) simplex projection, which characterizes the dimensionality of a system and the degree to which it appears to function deterministically; 2) S-maps, which quantify the degree of nonlinearity in a system; and 3) convergent cross-mapping, which offers a nonparametric approach to modeling causal effects . . . We conclude by discussing how EDM allows checking the assumptions of traditional model-based methods, such as residual autocorrelation tests, and we advocate for EDM because it does not assume linearity, stability, or equilibrium." (Li et al. 2021, Abstract)

Technicalities aside (see Li et al. 2021, for a detailed explanation), the following output (Figure F.4) demonstrates that we might be on the right track here. It seems that indeed DIGITALPROXY is carrying more information on logMARKETCAP than vice-versa. As shown, the experimental analysis suggests the likely causal direction is that an increase in DIGITALPROXY leads to a change in logMARKETCAP rate rather than the reverse case.

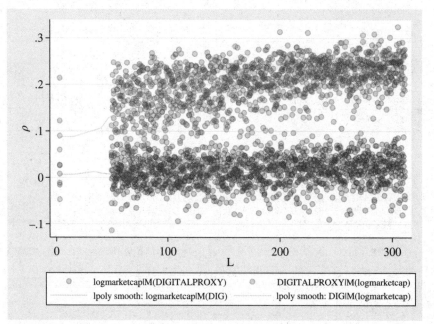

FIGURE F.4 EDM output (ρ – L convergence plot).

Do Industries Matter?

The following tables provide the detailed results of our interaction analysis (which in our main section we have just summarized as a simple table) and an additional test in which we analyze whether the differences between the chosen groups are statistically significant. The aforementioned complex analysis feeds into the * markers in the main section.

TABLE F.3   Industry category interaction effects.

Industry Category	Coeffi-cient	Pt	Lower 95% Confidence Interval	Higher 95% Confidence Interval	Overall Effect
DIGITAL-PROXY	−0.2829	0.0030	−0.4694	−0.0963	
Aerospace & Defense		(Base value)			
Automotive	0.5788	0.0000	0.3228	0.8349	0.2960
Banking	0.3325	0.0830	−0.0437	0.7086	0.0496
Chemicals	−0.4578	0.1370	−1.0621	0.1464	−0.7407
Computer Hardware	0.2173	0.0360	0.0139	0.4208	−0.0655
Computer Software & Services	0.3220	0.0020	0.1224	0.5216	0.0391
Conglomer-ates	0.4911	0.0010	0.2033	0.7790	0.2082
Consumer Durables	0.2215	0.0350	0.0161	0.4270	−0.0613
Consumer Nondurables	0.4302	0.0720	−0.0387	0.8990	0.1473
Diversified Services	0.3709	0.0000	0.1685	0.5733	0.0880
Drugs	0.4552	0.0010	0.1931	0.7173	0.1723
Electronics	0.2494	0.0240	0.0330	0.4657	−0.0335
Energy	0.2058	0.1590	−0.0812	0.4927	−0.0771

(continued)

TABLE F.3 (*continued*)

Industry Category	Coefficient	Pt	Lower 95% Confidence Interval	Higher 95% Confidence Interval	Overall Effect
Financial Services	0.5261	0.0000	0.2683	0.7840	0.2433
Food & Beverage	0.2205	0.1450	−0.0768	0.5177	−0.0624
Health Services	0.3352	0.0070	0.0913	0.5791	0.0523
Insurance	0.1473	0.5160	−0.2991	0.5937	−0.1356
Internet	0.2888	0.0060	0.0854	0.4922	0.0059
Leisure	0.4174	0.0010	0.1712	0.6636	0.1346
Manufacturing	0.4561	0.0000	0.2579	0.6543	0.1732
Materials & Construction	0.6074	0.0130	0.1311	1.0837	0.3245
Media	0.3103	0.0070	0.0876	0.5331	0.0275
Metals & Mining	0.3573	0.0010	0.1498	0.5648	0.0744
Real Estate	0.5441	0.0000	0.3143	0.7739	0.2612
Retail	0.1342	0.3810	−0.1669	0.4354	−0.1486
Specialty Retail	0.3622	0.0010	0.1577	0.5667	0.0794
Telecommunications	0.3099	0.0050	0.0938	0.5261	0.0271
Tobacco	−0.4681	0.0310	−0.8928	−0.0434	−0.7510
Transportation	0.6092	0.0060	0.1782	1.0402	0.3264
Utilities	0.4236	0.0290	0.0434	0.8038	0.1407
Wholesale	0.5126	0.0200	0.0828	0.9424	0.2297

Statistically significant for $p < 0.1$

TABLE F.4 Conditional marginal effects analysis.

Industry Category	Significant for DIGITALPROXY (0 to 30)
Aerospace & Defense	(Base value)
Automotive	Yes
Banking	Yes
Chemicals	No
Computer Hardware	Yes
Computer Software & Services	Yes
Conglomerates	Yes
Consumer Durables	Yes
Consumer NonDurables	Yes
Diversified Services	Yes
Drugs	Yes
Electronics	Yes
Energy	No
Financial Services	Yes
Food & Beverage	No
Health Services	Yes
Insurance	No
Internet	Yes
Leisure	Yes
Manufacturing	Yes
Materials & Construction	Yes
Media	Yes
Metals & Mining	Yes
Real Estate	Yes
Retail	No
Specialty Retail	Yes
Telecommunications	Yes
Tobacco	Yes
Transportation	Yes
Utilities	Yes
Wholesale	Yes

Yes: $p < 0.1$.

References

Andal-Ancion, Angela, Phillip A. Cartwright, and George S. Yip. 2003. "The Digital Transformation of Traditional Businesses." *MIT Sloan Management Review* 44 (4): 34–41.

Anderson, Mark C, Rajiv Banker, and Sury Ravindran. 2006. "Value Implications of Investments in Information Technology." *Management Science* 52 (9): 1359–1376.

Andriole, Stephen J. 2017. "Five Myths About Digital Transformation." *MIT Sloan Management Review* 58 (3): 20–22.

Aral, Sinan, and Peter Weill. 2007. "Information Technology Assets, Organizational Capabilities, and Firm Performance: How Resource Allocations and Organizational Differences Explain Performance Variation." *Organization Science* 18 (5): 763–780.

Baculard, Laurent-Pierre. 2017. "To Lead a Digital Transformation, CEOs Must Prioritize." *Harvard Business Review*, January 2.

Barlow, Jordan, Justin Giboney, Mark J. Keith, David Wilson, Ryan Schuetzler, Paul Benjamin Lowry, and Anthony Vance. 2011. "Overview and Guidance on Agile Development in Large Organizations." *Communications of the Association for Information Systems* 29: 25–44. doi: 10.2139/ssrn.1909431.

Barth, Mary E, William H Beaver, John RM Hand, and Wayne R Landsman. 1999. "Accruals, Cash Flows, and Equity Values." *Review of Accounting Studies* 4 (3–4): 205–229.

Baum, Christopher, and Mark Schaffer. 2021. "IVREG2H: Stata module to perform instrumental variables estimation using heteroskedasticity–based instruments." Statistical Software Components S457555, Department of Economics, Boston College.

Baumöl, Ulrike. 2016. "Die digitale Transformation und die erfolgsorientierte Unternehmenssteuerung–die Geschichte einer Revolution?" *Controlling* 28 (4–5): 230–234. doi: 10.15358/0935-0381-2016-4-5-230.

Becker, Jörg, Björn Niehaves, Jens Poeppelbuss, and Alexander Simons. 2010. "Maturity Models in IS Research." *ECIS Proceedings* 42.

Benson, David, and Rosemarie H. Ziedonis. 2009. "Corporate Venture Capital as a Window on New Technologies: Implications for the Performance of Corporate Investors When Acquiring Startups." *Organization Science* 20 (2): 329–351.

Berghaus, Sabine, and Andrea Back. 2016. "Stages in Digital Business Transformation: Results of an Empirical Maturity Study." *MCIS 2016 Proceedings*, 22.

Beswick, Paul. 2017. "Leading Digital Transformation Is Like Urban Planning." *Harvard Business Review*, August 2: 2–5.

Beutel, Sebastian. 2018. "The Relationship Between Digital Orientation and Firm Performance." *Proceedings of the International Conference on Information Systems—Bridging the Internet of People, Data and Things, ICIS 2018*, San Francisco, December 13–16.

Bharadwaj, Anandhi, Omar A. El Sawy, Paul A. Pavlou, and N. Venkatraman. 2013. "Digital Business Strategy: Toward a Next Generation of Insights." *MIS Quarterly* 37 (2): 471–482.

Bharadwaj, Anandhi S. 2000. "A Resource–Based Perspective on Information Technology Capability and Firm Performance: An Empirical Investigation." *MIS Quarterly* 24 (1): 169–169. doi: 10.2307/3250983.

Biddle, Gary C, Peter Chen, and Guochang Zhang. 2001. "When Capital Follows Profitability: Non-linear Residual Income Dynamics." *Review of Accounting Studies* 6 (2–3): 229–265.

Bock, Robert, Marco Iansiti, and Karim R. Lakhani 2017. "It Pays to Be a Digital Leader." *Harvard Business Review* 95 (3): 34.

Bohnsack, Rene, Andre Hanelt, David Marz, and Claudia Marante. 2018. "Same, Same, but Different!? A Systematic Review of the Literature on Digital Transformation." *Academy of Management Proceedings*, #16262.

Bojanowski, Piotr, Edouard Grave, Armand Joulin, and Tomas Mikolov. 2017. "Enriching Word Vectors with Subword Information." *Transactions of the Association for Computational Linguistics* 5: 135–146.

Briggs, Bill, and Buchholz, Scott. 2019. "Tech Trends 2019–Beyond the Digital Frontier." *Deloitte Insights 10th Anniversary Edition Tech Trends*. https://www2.deloitte .com/content/dam/Deloitte/br/Documents/technology/DI_TechTrends2019.pdf.

Brockhoff, Klaus K. 1999. "Technological Progress and the Market Value of Firms." International *Journal of Management Reviews* 1 (4): 485.

Brynjolfsson, Erik, and Lorin M. Hitt. 2000. "Beyond Computation: Information Technology, Organizational Transformation and Business Performance." *Journal of Economic Perspectives* 14 (4): 23–48.

Brynjolfsson, Erik, and Andrew McAfee. 2014. *The Second Machine Age: Work, Progress, and Prosperity in a Time of Brilliant Technologies*. New York: W.W. Norton.

Buchholz, Scott, and Bill Briggs. 2022. "Deloitte Tech Trends." Accessed March 18, 2022, at https://www2.deloitte.com/us/en/insights/focus/tech–trends.html.

Bughin, Jacques, and Tanguy Catlin. 2017. "What successful digital transformations have in common." *Harvard Business Review Digital Articles*: 1–5.

Cao, Lan, Kannan Mohan, Peng Xu, and Balasubramaniam Ramesh. 2009. "A framework for adapting agile development methodologies." *European Journal of Information Systems* 18 (4): 332–343. doi: 10.1057/ejis.2009.26.

Callen, Jeffrey L, and Dan Segal. 2005. "Empirical Tests of the Feltham–Ohlson (1995) Model." *Review of Accounting Studies* 10 (4): 409–429.

Chaffey, Dave. 2010. "Applying Organisational Capability Models to Assess the Maturity of Digital-Marketing Governance." *Journal of Marketing Management* 26 (3–4): 187–196. doi: 10.1080/02672571003612192.

Charan, Ram. 2016. "How to Transform a Traditional Giant into a Digital One." *Harvard Business Review*, February 26.

Chen, Wilbur, and Suraj Srinivasan. 2019. "Going Digital: Implications for Firm Value and Performance." Working Paper, Harvard Business School.

Choi, Young-Soo, John O'Hanlon, and Peter Pope. 2003. "Linear Information Models in Residual Income-Based Valuation: Intercept Parameters and Valuation Bias." Working Paper, Lancaster University.

Chouliaras, Andreas. 2015. "The Pessimism Factor: SEC EDGAR Form 10-k Textual Analysis and Stock Returns." Available at SSRN 2627037.

Cinelli, Carlos, Jeremy Ferwerda, and Chad Hazlett. 2020. "Sensemakr: Sensitivity Analysis Tools for OLS in R and Stata." Available at SSRN 3588978.

Clippinger, Richard F. 1955. "ECONOMICS of the Digital Computer." *Harvard Business Review* 33 (1): 77–88.

Cohen, Lauren, Christopher Malloy, and Quoc Nguyen. 2020. "Lazy Prices." *The Journal of Finance* 75 (3): 1371–1415.

Copeland, Thomas E., and Vladimir Antikarov. 2001. *Real Options: A Practitioner's Guide*. New York: Texere.

Damodaran, Aswath. 2013. "Living with Noise: Valuation in the Face of Uncertainty." *Journal of Applied Finance* 23 (2): 6–22.

Damodaran, Aswath. 2017. *Narrative and Numbers: The Value of Stories in Business*. New York: Columbia Business School Publishing.

Davenport, Thomas H., and George Westerman. 2018. "Why So Many High-Profile Digital Transformations Fail." *Harvard Business Review*, March 9.

Dechow, Patricia M, Amy P Hutton, and Richard G Sloan. 1999. "An Empirical Assessment of the Residual Income Valuation Model." *Journal of Accounting Economics* 26 (1–3): 1–34.

Dehning, Bruce, Vernon J. Richardson, Andrew Urbaczewski, and John D. Wells. 2004. "Reexamining the Value Relevance of E-Commerce Initiatives." *Journal of Management Information Systems* 21 (1): 55–82. doi: 10.1080/07421222.2004.11045788.

Dehning, Bruce, Vernon J. Richardson, and Robert W. Zmud. 2003. "The Value Relevance of Announcements of Transformational Information Technology Investments." *MIS Quarterly* 27 (4): 637–656. doi: 10.2307/30036551.

Deloitte. 2018. "Deloitte Millennial Survey: Millennials Disappointed in Business, Unprepared for Industry 4.0." https://www2.deloitte.com/content/dam/Deloitte/global/Documents/About-Deloitte/gx-2018-millennial-survey-report.pdf.

Deloitte. 2019. "The Deloitte Global Millennial Survey: Societal Discord and Technological Transformation Create a 'Generation Disrupted.'" https://www2.deloitte.com/content/dam/Deloitte/global/Documents/About-Deloitte/deloitte-2019-millennial-survey.pdf.

Devaraj, Sarv, and Rajiv Kohli. 2002. *The Information Technology Payoff: Measuring the Business Value of Information Technology Investments*. New York: Financial Times, Prentice Hall.

ENISA. 2012. "Introduction to Return on Security Investment." Accessed January 12, 2022, at https://www.enisa.europa.eu/publications/introduction-to-return-on-security-investment.

Falkum, A. 2011. "Rechnungswesenorientierte Unternehmensbewertung: Einsatz und Eignung der kennzahlenorientierten Fundamentalanalyse bei der Erweiterung LIM-gestützter Bewertungsverfahren." PhD, Wirtschaftswissenschaftliche Fakultät, Universität Würzburg.

Fama, Eugene F. 1970. "Efficient Capital Markets: A Review of Theory and Empirical Work." *The Journal of Finance* 25 (2): 383–417. doi: 10.2307/2325486.

Fichman, Robert G., Mark Keil, and Amrit Tiwana. 2005. "Beyond Valuation: 'Options Thinking' in Information Technology Project Management." *California Management Review* 47 (2): 74–96.

Fitzgerald, Michael, Nina Kruschwitz, Didier Bonnet, and Michael Welch. 2013. "Embracing Digital Technology: A New Strategic Imperative." *MIT Sloan Management Review*, October 7.

Foutty, Janet, and Mike Bechtel. 2022. "What's all the buzz about the metaverse?" Deloitte Perspectives. Retrieved from https://www2.deloitte.com/us/en/pages/center-for-board-effectiveness/articles/whats-all-the-buzz-about-the-metaverse.html.

Gale, Michael, and Chris Aarons. 2017. *The Digital Helix: Transforming Your Organization's DNA to Thrive in the Digital Age*. Austin: Greenleaf Book Group.

Galindo-Martín, Miguel-Ángel, María-Soledad Castaño-Martínez, and María-Teresa Méndez–Picazo. 2019. "Digital Transformation, Digital Dividends and Entrepreneurship: A Quantitative Analysis." *Journal of Business Research* 101: 522–527. doi: https://doi.org/10.1016/j.jbusres.2018.12.014.

Gao, Zhan, James N. Myers, Linda A. Myers, and Wan-Ting Wu. 2019. "Can a Hybrid Method Improve Equity Valuation? An Empirical Evaluation of the Ohlson and Johannesson (2016) Model." *The Accounting Review* 94 (6): 227–252.

Gartner. 2018. "Every Organizational Function Needs to Work on Digital Transformation." *Harvard Business Review*, November 27.

Gassmann, Oliver, Karolin Frankenberger, and Michaela Csik. 2014. *The Business Model Navigator: 55 Models That Will Revolutionise Your Business*. Harlow, UK: Pearson.

Gimpel, H. and Röglinger, M. 2015. "Digital Transformation: Changes and Chances—Insights Based on Empirical Study." Working Paper, Universität Bayreuth.

Golden, Deborah, and Vikram Kunchala. 2021. "An Integrated Cyber Approach to Your Cloud Migration Strategy." Accessed March 28, 2022 at https://www2.deloitte.com/us/en/insights/topics/digital–transformation/integrated–cyber–security–cloud–migration--strategy.html.

Gong, Cheng, and Vincent Ribiere. 2021. "Developing a Unified Definition of Digital Transformation." *Technovation* 102 (April): 102217.

Gray, Paul, Omar A. El Sawy, Guillermo Asper, and Magnus Thordarson. 2013. "Realizing Strategic Value Through Center-Edge Digital Transformation in Consumer-Centric Industries." *MIS Quarterly Executive* 12 (1): 1–17.

Haffke, Ingmar, Bradley Kalgovas, and Alexander Benlian. 2017. "Options for Transforming the Information Technology Function Using Bimodal Information Technology." *MIS Quarterly Executive* 16 (2): 101–120.

Hamel, Stéphane. 2009. "The Web analytics maturity model." Accessed January 24, 2021, at https://www.stephanehamel.net/WAMM_ShortPaper_091017.pdf.

Haritash, Rohit. 2018. "EDGAR-reports-Text-Analysis." Accessed March 12, 2020, at https://github.com/rohitharitash/EDGAR-reports-Text-Analysis/blob/master/README.md.

Henderson, Rebecca M., and Kim B. Clark. 1990. "Architectural Innovation: The Reconfiguration of Existing Product Technologies and the Failure of Established Firms." *Administrative Science Quarterly* 35 (1): 9–30.

Henriette, Emily, Mondher Feki, and Imed Boughzala. 2015. "The Shape of Digital Transformation: A Systematic Literature Review." *MCIS 2015 Proceedings*, 10.

Hess, Thomas, Christian Matt, Alexander Benlian, and Florian Wiesböck. 2016. "Options for Formulating a Digital Transformation Strategy." *MIS Quarterly Executive* 15 (2): 123–139.

Hinings, Bob, Thomas Gegenhuber, and Royston Greenwood. 2018. "Digital Innovation and Transformation: An Institutional Perspective." *Information and Organization* 28 (1): 52–61. doi: https://doi.org/10.1016/j.infoandorg.2018.02.004.

Hitt, Lorin M., and Erik Brynjolfsson. 1996. "Productivity, Business Profitability, and Consumer Surplus: Three Different Measures of Information Technology Value." *MIS Quarterly* 20 (2): 121–142.

Hossnofsky, Verena, and Sebastian Junge. 2019. "Does the Market Reward Digitalization Efforts? Evidence from Securities Analysts' Investment Recommendations." *Journal of Business Economics* 89 (8): 965–994.

Hund, Axel, Katharina Drechsler, and Victoria Alexandra Reibenspiess. 2019. "The Current State and Future Opportunities of Digital Innovation: A Literature Review." *ECIS Proceedings 2019*.

Intrinio. 2020. "US company fundamentals." Accessed December 20, 2020, at https://product.intrinio.com/financial-data/us-fundamentals-financials-metrics-ratios.

Kane, Gerald. 2019. *The Technology Fallacy: People Are the Real Key to Digital Transformation*. London: Taylor & Francis.

Kane, Gerald C., Doug Palmer, David Kiron, and Natasha Buckley. 2015. "Strategy, Not Technology, Drives Digital Transformation." *MIT Sloan Management Review* and Deloitte University Press, July 14.

Kane, Gerald C., Doug Palmer, David Kiron, and Natasha Buckley. 2018. "Coming of Age Digitally." *MIT Sloan Management Review* and Deloitte Insights, June 15.

Kane, Gerald C., Doug Palmer, Anh Nguyen Phillips, David Kiron, and Natasha Buckley. 2016. "Aligning the Organization for Its Digital Future." *MIT Sloan Management Review* 58 (1): 1–28.

Kane, Gerald C., Doug Palmer, Anh Nguyen Phillips, David Kiron, and Natasha Buckley. 2017. "Achieving Digital Maturity." *MIT Sloan Management Review* 59 (1): 1–29.

Kawohl, Julian, and Patrick Hüpel. 2018. "Digitale Transformation: Wie weit sind die DAX 30?" Accessed January 10, 2019, at https://docs.wixstatic.com/ugd/63eb59_42c1bc11e47544309d2c4a357e4099cd.pdf.

Keil, Thomas, Ruth Gunther McGrath, and Taina Tukiainen. 2009. "Gems from the Ashes: Capability Creation and Transformation in Internal Corporate Venturing." *Organization Science* 20 (3): 601–620.

Kohli, Rajiv, and Sarv Devaraj. 2003. "Measuring Information Technology Payoff: A Meta-Analysis of Structural Variables in Firm-Level Empirical Research." *Information Systems Research* 14 (2): 127–145. doi: 10.1287/isre.14.2.127.16019.

Kohli, Rajiv, Sarv Devaraj, and Terence T. Ow. 2012. "Does Information Technology Investment Influence a Firm's Market Value? A Case of Non-publicly Traded Healthcare Firms." *MIS Quarterly* 36 (4): 1145–1163. doi: 10.2307/41703502.

Kohli, Rajiv, and Varun Grover. 2008. "Business Value of Information Technology: An Essay on Expanding Research Directions to Keep up with the Times." *Journal of the Association for Information Systems* 9 (1): 23–39.

Lafley, A. G., and Roger L. Martin. 2013. *Playing to Win: How Strategy Really Works*. Brighton: Harvard Business Press.

Leonardi, Paul M. 2007. "Activating the Informational Capabilities of Information Technology for Organizational Change." *Organization Science* 18 (5): 813–831.

Li, Feng. 2006. "Do Stock Market Investors Understand the Risk Sentiment of Corporate Annual Reports?" Available at SSRN 898181.

Li, Jinjing, Michael J. Zyphur, George Sugihara, and Patrick J. Laub. 2021. "Beyond Linearity, Stability, and Equilibrium: The Edm Package for Empirical Dynamic Modeling and Convergent Cross-mapping in Stata." *The Stata Journal* 21 (1): 220–258.

Lo, Kin, and Thomas Lys. 2000. "The Ohlson Model: Contribution to Valuation Theory, Limitations, and Empirical Applications." *Journal of Accounting, Auditing, Finance* 15 (3): 337–367.

Loria, Steven. 2018. "textblob Documentation Release 0.15." Accessed October 12, 2021.

Loughran, Tim, and Bill McDonald. 2011. "When Is a Liability Not a Liability? Textual Analysis, Dictionaries, and 10-Ks." *The Journal of Finance* 66 (1): 35–65.

Magnusson, Johan, and Bendik Bygstad. 2014. "Technology Debt: Toward a New Theory of Technology Heritage," in *Proceedings of the 22nd European Conference on Information Systems*, Tel Aviv.

Mani, Deepa, Anand Nandkumar, and Anandhi Bharadwaj. 2018. "Market Returns to Digital Innovations: A Group Based Trajectory Approach." *SSRN Electronic Journal.* doi: 10.2139/ssrn.3241483

Martin, Roger L. 2021. "Playing to Win Practitioner Insights." Accessed February 20, 2022, at https://rogermartin.medium.com/strategy-transformation-4c8d453d40bc.

Mazzone, Dominic. 2014. *Digital or Death: Digital Transformation— The Only Choice for Business to Survive, Smash or Conquer.* Mississauga, Ontario: Smashbox Consulting.

McAfee, Andres, and Erik Brynjolfsson. 2008. "Investing in the Information Technology That Makes a Competitive Difference." *Harvard Business Review* 86 (7/8): 98–107.

McCrae, Michael, and Henrik Nilsson. 2001. "The Explanatory and Predictive Power of Different Specifications of the Ohlson (1995) Valuation Models." *European Accounting Review* 10 (2): 315–341.

Mencken, H. L. 1920. Chapter 4: "The Divine Afflatus, Start." In *Prejudices: Second Series.* New York: Alfred A. Knopf.

Mintzberg, Henry, Bruce W. Ahlstrand, and Joseph Lampel. 1998. *Strategy Safari: The Complete Guide Through the Wilds of Strategic Management.* London: Financial Times/Prentice Hall. Book.

Mithas, Sunil, Narayan Ramasubbu, and V. Sambamurthy. 2011. "How Information Management Capability Influences Firm Performance." *MIS Quarterly* 35 (1): 237–256.

Morgan, Steve. 2020. "Cybercrime to Cost the World $10.5 Trillion Annually by 2025." https://cybersecurityventures.com/cybercrime-damages-6-trillion-by-2021/.

Morone, Joseph. 1989. "Strategic Use of Technology." *California Management Review* 31 (4): 91–110.

Morton, Josh, Patrick Stacey, and Matthias Mohn. 2018. "Building and Maintaining Strategic Agility: An Agenda and Framework for Executive Information Technology Leaders." *California Management Review* 61 (1): 94–113. doi: 10.1177/0008125618790245.

Muhanna, Waleed A., and M. Dale Stoel. 2010. "How Do Investors Value Information Technology? An Empirical Investigation of the Value Relevance of Information Technology Capability and Information Technology Spending Across Industries." *Journal of Information Systems* 24 (1): 43–66. doi: 10.2308/jis.2010.24.1.43.

Mullins, John, and Randy Komisar. 2011. "Measuring Up: Dashboarding for Innovators." *Business Strategy Review* 22 (1): 7–16. doi: 10.1111/j.1467–8616.2011.00723.x.

Myers, James N. 2000. "Discussion: 'The Feltham-Ohlson (1995) Model: Empirical Implications.'" *Journal of Accounting, Auditing, Finance* 15 (3): 332–335.

Nambisan, Satish, Kalle Lyytinen, Ann Majchrzak, and Michael Song. 2017. "Digital Innovation Management: Reinventing Innovation Management Research in a Digital World." *MIS Quarterly* 41 (1): 223–238.

Nanda, Rich, Ragu Gurumurthy, Sam Roddick, and Deborah Golden. 2021a. "A New Language for Digital Transformation." Deloitte, September 23. https://www2.deloitte.com/us/en/insights/topics/digital--transformation/digital--transformation--approach.html.

Nanda, Rich, Ben Philips, and Bill Jarmuz. 2021b. "The Exponential Enterprise." Accessed March 20, 2022, at https://www2.deloitte.com/us/en/pages/consulting/articles/exponential–enterprise.html.

Nazarov, D. M., E. K. Fitina, and A. O. Juraeva. 2019. "Digital Economy as a Result of the Genesis of the Information Revolution of Society." *MTDE Proceedings*, Atlantis Press, Yekatarinburg.

NLTK. 2020. Accessed May 5, 2020, at https://www.nltk.org/.

Ohlson, James A. 1995. "Earnings, Book Values, and Dividends in Equity Valuation." *Contemporary Accounting Research* 11 (2): 661–687.

Ohlson, James A. 2001. "Earnings, Book Values, and Dividends in Equity Valuation: An Empirical Perspective." *Contemporary Accounting Research* 18 (1): 107–120.

Oprescu, Miruna, Vasilis Syrgkanis, Keith Battocchi, Maggie Hei, and Greg Lewis. 2019. "EconML: A Machine Learning Library for Estimating Heterogeneous Treatment Effects." 33rd Conference on Neural Information Processing Systems (NEURIPS 2019), Vancouver, Canada. https://cpb-us-w2.wpmucdn.com/sites.coecis.cornell.edu/dist/a/238/files/2019/12/Id_112_final.pdf.

Osmundsen, Karen, Jon Iden, and Bendik Bygstad. 2018. "Digital Transformation: Drivers, Success Factors, and Implications." *MCIS Proceedings* 37. Corfu.

Ota, Koji. 2002. "A Test of the Ohlson (1995) Model: Empirical Evidence from Japan." *The International Journal of Accounting* 37 (2): 157–182.

Parida, Vinit, David Sjödin, and Wiebke Reim. 2019. "Reviewing Literature on Digitalization, Business Model Innovation, and Sustainable Industry: Past Achievements and Future Promises." *Sustainability* 11 (2): 1–18.

Patel, Ajay, Alexander Sands, Chris Callison-Burch, and Marianna Apidianaki. 2018. "Magnitude: A Fast, Efficient Universal Vector Embedding Utility Package." *arXiv preprint arXiv:1810.11190.*

Peppard, Joe, and John Ward. 2005. "Unlocking Sustained Business Value from Information Technology Investments." *California Management Review* 48 (1): 52–70.

Peters, Matthew E, Mark Neumann, Mohit Iyyer, Matt Gardner, Christopher Clark, Kenton Lee, and Luke Zettlemoyer. 2018. "Deep Contextualized Word Representations." *arXiv preprint arXiv:.05365.*

Pramanik, Himadri Sikhar, Manish Kirtania, and Ashis K. Pani. 2019. "Essence of Digital Transformation—Manifestations at Large Financial Institutions from North America." *Future Generation Computer Systems*: 323–343. doi: https://doi.org/10.1016/j.future.2018.12.003.

Proença, Diogo, and José Borbinha. 2016. "Maturity Models for Information Systems—A State of the Art." *Procedia Computer Science* 100: 1042–1049. doi: https://doi.org/10.1016/j.procs.2016.09.279.

Python. 2020. Accessed January 12, 2020, at https://www.python.org/.

Raskino, Mark, and Graham Waller. 2015. *Digital to the Core: Remastering Leadership for Your Industry, Your Enterprise and Yourself.* Routledge.

Remane, Gerrit, Andre Hanelt, Florian Wiesböck, and Lutz Kolbe. 2017. "Digital Maturity in Traditional Industries: An Exploratory Analysis." *ECIS Proceedings 2018.*

Roberts, Nicholas, and Varun Grover. 2012. "Leveraging Information Technology Infrastructure to Facilitate a Firm's Customer Agility and Competitive Activity: An Empirical Investigation." *Journal of Management Information Systems* 28 (4): 231–270. doi: 10.2753/MIS0742-1222280409.

Rodriguez-Ramos, Jaime. 2018. *Beyond Digital —Six Exponential Revolutions That Will Change Our World.* Madrid: Amazon Media.

Rogers, David L. 2016. *The Digital Transformation Playbook: Rethink Your Business for the Digital Age.* New York: Columbia University Press.

Sabherwal, Rajiv, and Anand Jeyaraj. 2015. "Information Technology Impacts on Firm Performance: An Extension of Kohli and Devaraj (2003)." *MIS Quarterly* 39 (4): 809–836.

Saldanha, Tony. 2019. *Why Digital Transformations Fail: The Surprising Disciplines of How to Take Off and Stay Ahead:* Berrett–Koehler Publishers.

Salomo, Sören, Katrin Talke, and Nanja Strecker. 2008. "Innovation Field Orientation and Its Effect on Innovativeness and Firm Performance." *Journal of Product Innovation Management* 25 (6): 560–576. doi: 10.1111/j.1540–5885.2008.00322.x.

Salviotti, Gianluca. 2022. "Rediscovering the Fundamentals of Value Creation." In Gianluigi Castelli, Severino Meregalli, and Ferdinando Pennarola (Eds.), *The Post-Digital Enterprise: Going Beyond the Hype* (p. 44). Cham: Springer.

Saunders, Adam, and Erik Brynjolfsson. 2016. "Valuing Information Technology Related Intangible Assets." *MIS Quarterly* 40 (1): 83–110.

Schallmo, Daniel, Andreas Rusnjak, Johanna Anzengruber Thomas Werani, and Michael Jünger. 2017. *Digitale Transformation von Geschäftsmodellen: Grundlagen, Instrumente und Best Practices.* Wiesbaden: Springer Gabler.

Schneider, Robin. 2018. "Valuing Investments in Digital Business Transformation: A Real Options Approach." Accessed January 5, 2019 at http://www.realoptions.org/openconf2018/data/papers/162.pdf.

Schönbohm, Avo, and Ulrich Egle. 2017. "Controlling der Digitalen Transformation." In *Digitale Transformation von Geschäftsmodellen: Grundlagen, Instrumente und Best Practices,* edited by Daniel Schallmo, Andreas Rusnjak, Johanna Anzengruber, Thomas Werani and Michael Jünger, 213–236. Wiesbaden: Springer Fachmedien.

Schwab, Klaus. 2017. *The Fourth Industrial Revolution.* New York: Crown Publishing Group.

Scott, Susan V., John Van Reenen, and Markos Zachariadis. 2017. "The Long-term Effect of Digital Innovation on Bank Performance: An Empirical Study of SWIFT Adoption in Financial Services." *Research Policy* 46 (5): 984–1004. doi: https://doi.org/10.1016/j.respol.2017.03.010.

Sebastian, Ina M., Jeanne W. Ross, Cynthia Beath, Martin Mocker, Kate G. Moloney, and Nils O. Fonstad. 2017. "How Big Old Companies Navigate Digital Transformation." *MIS Quarterly Executive*: 197–213.

SEC. 2020a. "Accessing EDGAR Data." U.S. Securities and Exchange Commission. Accessed January 2, 2020, at https://www.sec.gov/edgar/searchedgar/accessing-edgar-data.htmhbh.

SEC. 2020b. "Descriptions of SEC Forms." U.S. Securities and Exchange Commission. Accessed January 2, 2020, at https://www.sec.gov/info/edgar/forms/edgform.pdf.

SEC. 2020c. "Financial Reporting Manual." U.S. Securities and Exchange Commission. Accessed January 2, 2020, at https://www.sec.gov/divisions/corpfin/cffinancialreportingmanual.pdf.

Sharma, Amit, and Emre Kiciman. 2020. "DoWhy: An End-to-End Library for Causal Inference." *arXiv preprint arXiv: 2011.04216.*

Sircar, Sumit, Joe L. Turnbow, and Bijoy Bordoloi. 2000. "A Framework for Assessing the Relationship Between Information Technology Investments and Firm Performance." *Journal of Management Information Systems* 16 (4): 69–97.

Sjödin, David R., Vinit Parida, Markus Leksell, and Aleksandar Petrovic. 2018. "Smart Factory Implementation and Process Innovation." *Research–Technology Management* 61 (5): 22–31. doi: 10.1080/08956308.2018.1471277.

Smit, Han T. J., and Lenos Trigeorgis. 2009. "Valuing Infrastructure Investment: AN OPTION GAMES APPROACH." *California Management Review* 51 (2): 79–100.

spaCy. 2020. Accessed May 5, 2020, at https://spacy.io/.

STATA. 2020. "Nonparametric Series Regression." Accessed April 10, 2020, at https://www.stata.com/new-in-stata/nonparametric-series-regression/.

Strecker, Nanja. 2009. *Innovation Strategy and Firm Performance: An Empirical Study of Publicly Listed Firms.* Wiesbaden: Gabler.

Subramani, Mani, and Eric Walden. 2001. "The Impact of E-Commerce Announcements on the Market Value of Firms." *Information Systems Research* 12 (2): 135.

Szutowski, Dawid, and Julia Szułczyńska. 2017. "The Model Approach to Linking Innovation Announcements and Market Value of Equity in Service Sector." *Wpływ Ogłoszeń O Innowacjach Na Wartość Dla Akcjonariuszy Przedsiębiorstw Sektora Usług Rynkowych. Ujęcie Modelowe* (478): 425–435. doi: 10.15611/pn.2017.478.39.

Templeton, Gary F., and Laurie L. Burney. 2017. "Using a Two-Step Transformation to Address Non-Normality from a Business Value of Information Technology Perspective." *Journal of Information Systems* 31 (2): 149–164. doi: 10.2308/isys-51510.

Tilson, David, Kalle Lyytinen, and Carsten Sørensen. 2010. "Digital Infrastructures: The Missing IS Research Agenda." Research commentary. *Information Systems Research* 21 (4): 748–759.

UND. 2020. "University of Notre Dame Software Repository for Accounting and Finance: Loughran–McDonald Sentiment Word Lists." Accessed April 4, 2019.

Vartanian, Vatchagan. 2003. *Innovationsleistung und Unternehmenswert: Empirische Analyse wachstumsorientierter Kapitalmärkte.* Wiesbaden: Springer-Verlag.

Venkatesh, Ramamurthy, and Tarun Singhal. 2019. "Conflating the Dimensions of Business Innovation and Digital Transformation." *Amity Global Business Review* 47: 57–63.

Verhoef, Peter C., and Tammo H. A. Bijmolt. 2019. "Marketing Perspectives on Digital Business Models: A Framework and Overview of the Special Issue." *International Journal of Research in Marketing* 36 (3): 341–349.

Verhoef, Peter C., Thijs Broekhuizen, Yakov Bart, Abhi Bhattacharya, John Qi Dong, Nicolai Fabian, and Michael Haenlein. 2019. "Digital Transformation: A Multidisciplinary Reflection and Research Agenda." *Journal of Business Research* 122: 889–901.

Vukšić, Vesna Bosilj, Lucija Ivančić, and Dalia Suša Vugec. 2018. "A Preliminary Literature Review of Digital Transformation Case Studies." ICMIT 2018: 20th International Conference on Managing Information Technology.

Walker, Richard, and Brian Hansen. 2021. "Deloitte's 2021 Global Blockchain Survey." Accessed March 22, 2022, at https://www2.deloitte.com/us/en/insights/topics/understanding-blockchain-potential/global-blockchain-survey.html.

Wang, Sarah, and Martin Casado. 2021. "The Cost of Cloud, a Trillion Dollar Paradox." Accessed December 8, 2021, at https://a16z.com/2021/05/27/cost-of-cloud-paradox-market-cap-cloud-lifecycle-scale-growth-repatriation-optimization/.

Weill, Peter, and Stephanie L. Woerner. 2018. "Is Your Company Ready for a Digital Future?" *MIT Sloan Management Review* 59 (2): 21–25.

Westerman, George, and Didier Bonnet. 2015. "Revamping Your Business Through Digital Transformation." *MIT Sloan Management Review* 56 (3): 10–13.

Westerman, George, Didier Bonnet, and Andrew McAfee. 2014. *Leading Digital: Turning Technology into Business Transformation.* Brighton: Harvard Business Press.

Westerman, George, Claire Calméjane, Didier Bonnet, Patrick Ferraris, and Andrew McAfee. 2011. "Digital Transformation: A Roadmap for Billion-Dollar Organizations." MIT Center for Digital Business and Capgemini Consulting.

Westerman, George, Maël Tannou, Didier Bonnet, Patrick Ferraris, and Andrew McAfee. 2012. "The Digital Advantage: How Digital Leaders Outperform Their Peers in Every Industry." MIT Sloan Management and Capgemini Consulting.

White, Samantha. 2016. "Intangibles Drive Value in the Digital Age." *Journal of Accountancy* 221 (3): 21–21.

Williams, Christopher, and Daniel Schallmo. 2018. "History of Digital Transformation." In *Digital Transformation Now!: Guiding the Successful Transformation of Your Business Model*, 3–8. Wiesbaden: Springer.

WorldBank. 2018. "World Development Report 2019: The Changing Nature of Work." Accessed March 3, 2020, at https://elibrary.worldbank.org/doi/abs/10.1596/978-1-4648-1328-3_fm.

Wroblewski, Julian. 2018. "Digitalization and Firm Performance: Are Digitally Mature Firms Outperforming Their Peers?" MsC, Lund University.

yahoo. 2020. "yahoo!finance." Accessed October 10, 2020 at https://finance .yahoo.com/.

Youngjin, Yoo, Richard J. Boland Jr, Kalle Lyytinen, and Ann Majchrzak. 2012. "Organizing for Innovation in the Digitized World." *Organization Science* 23 (5): 1398–1408. doi: 10.1287/orsc.1120.0771.

Yunis, Manal, Abbas Tarhini, and Abdulnasser Kassar. 2018. "The Role of ICT and Innovation in Enhancing Organizational Performance: The Catalysing Effect of Corporate Entrepreneurship." *Journal of Business Research* 88: 344–356. doi: https://doi .org/10.1016/j.jbusres.2017.12.030.

Zomer, Thayla, Andy Neely, and Veronica Martinez. 2018. "Enabling Digital Transformation: An Analysis Framework." Working Paper. Cambridge University.

About the Author

Tim Bottke is an Associate Professor for Strategy and Digital Transformation at SDA Bocconi, a *Financial Times/Forbes/Bloomberg Businessweek* top-five European business school, and a Senior Strategy Partner at Monitor Deloitte. He has more than 22 years of top management consulting and (digital) transformation experience from two global strategy boutiques and Deloitte, working with clients in more than 20 countries.

Index